SpringerBriefs in Molecular Science

Green Chemistry for Sustainability

Series Editor

Sanjay K. Sharma

For further volumes:
http://www.springer.com/series/10045

Ta Yeong Wu · Ningqun Guo
Chee Yang Teh · Jacqueline Xiao Wen Hay

Advances in Ultrasound Technology for Environmental Remediation

 Springer

Ta Yeong Wu
School of Engineering
Monash University
Selangor Darul Ehsan
Malaysia

Ningqun Guo
School of Engineering
Monash University
Selangor Darul Ehsan
Malaysia

Chee Yang Teh
School of Engineering
Monash University
Selangor Darul Ehsan
Malaysia

Jacqueline Xiao Wen Hay
School of Engineering
Monash University
Selangor Darul Ehsan
Malaysia

ISSN 2212-9898
ISBN 978-94-007-5532-1   ISBN 978-94-007-5533-8   (eBook)
DOI 10.1007/978-94-007-5533-8
Springer Dordrecht Heidelberg New York London

Library of Congress Control Number: 2012947061

Printed on acid-free paper

Springer is part of Springer Science+Business Media (www.springer.com)

Contents

Chapter 1
Introduction

Abstract Removal of toxic and organic conatminants has become a matter of considerable interest. If the untreated contaminant is released into environments, it is certain to cause significant environmental problems due to its accumulation in soil and water environments. Although biological method is generally used in the treatment of contaminants, this method is not perfect and continuous efforts to improve biological remediation are necessary. On the other hand, ultrasound technology could become an alternative method for waste management and environmental remediation. Contrary to conventional treatments, ultrasound technology may offer advantages, such as environmentally friendly, low costs, compact and others. The application of ultrasound technology for environmental remediation is still in developing stage but it is growing rapidly and holds a promising future as one of the leading "green" technologies for environmental remediation.

Keywords Biological treatment · Contaminants · Environmental remediation · Ultrasound technology · Waste management

1.1 Introduction

Any unwanted substance introduced into the environment is referred to as a 'contaminant'. Deleterious effects or damages by the introduction of contaminants into environments will lead to 'pollution', a process by which a resource (natural or man-made) is rendered unfit for use, more often than not, by humans (Megharaj et al. 2011). It is especially true when a wide range of toxic contaminants are produced as by-products from the industries which synthesize new products such as agricultural chemicals, pharmaceuticals, dyes, polymers, and others. Numerous

T. Y. Wu et al., *Advances in Ultrasound Technology for Environmental Remediation*,
SpringerBriefs in Green Chemistry for Sustainability, DOI: 10.1007/978-94-007-5533-8_1,
© The Author(s) 2013

industrial contaminants are now present in the terrestrial environment and these contaminants may accumulate in soil and water environments. The increasing public pressure and scientific concern led most countries to develop biological pathways/treatments for the protection of ecosystems (Ghasemi et al. 2011).

Biological treatment is extensively used in waste management because of its low cost and the fact that it does not require any chemicals other than nutrients (Thangavadivel et al. 2012). However, it is generally a slow process, does not withstand shock loading and is not efficient at very high concentrations of organic pollutants or when the pollutants are combined with other toxic contaminants (Thangavadivel et al. 2012). Also, biological treatment of wastewater generates waste activated sludge, which is difficult to be digested due to rate-limiting cell lysis (Khanal et al. 2007). Moreover, the potential to degrade organic contaminants varies among microbial group of species and is dose-dependent (Megharaj et al. 2011). Although efforts have been directed toward the development of other remediation technologies on such contaminants, many contaminated sites are still remained untreated either due to the prohibitive cost of remediation or the lack of technologies that can clean up the environment to levels required by the regulators. Also, the proposed remediation technologies are usually energy-intensive, may produce their own pollutant emissions, often require many years for implementation with long-term monitoring and sometimes even cause controversy in neighboring communities (Holland 2011). Thus, it is absolutely imperative to continue finding and developing a more environmentally and sustainable remediation technology.

Ultrasound refers to an inaudible-cyclic-sound-pressure wave with frequencies in the range of 0.02–500 MHz, greater than the upper limit of human hearing. Ultrasound has been used for diverse purposes in many different areas. Of late, the application of ultrasonic technology has been receiving wide attention as a green technology (Estager 2012) in water/wastewater treatment and environmental remediation. Ultrasound technology uses acoustic cavitation to achieve physical as well as chemical effects within the solution, which help degrade pollutants (Thangavadivel et al. 2012). According to Chen (2012), ultrasound technology was attempted in the degradation of recalcitrant organic pollutants in aqueous phase, decontamination of sediments, assistance of membrane filtration for membrane cleaning to reduce fouling, disinfection, and others. The applications of ultrasound technology in environmental remediation hold a promising future. Contrary to conventional methods, ultrasound technology may offer advantages such as environmentally friendly (because no toxic chemicals are used or produced), low costs (in small-scale basis) and compact (because the method allows on-site treatment) (Pham et al. 2009). Also, ultrasound technology operates in normal atmospheric conditions, does not generate sludge and is easy to be installed and operated (Thangavadivel et al. 2012).

Ultrasound technology alone is generally not feasible to be used in large-scale treatment process because it requires costly equipment and consumes high amount of energy, and not all of the cavitational energy can be transformed into chemical and physical effects (Pang et al. 2011). These weaknesses result in limitation of ultrasound technology to be used widely in a real wastewater treatment plant. Ultrasonic applications in environmental areas are still in developing stage but they are growing rapidly and receiving much attention among the environmentalists and engineers. This is because the high operating cost in large scale operation could be partially off-set by operating ultrasound technology at milder and optimum operation conditions, which still enhance the degradation rate, reduce the reaction time, and eliminate the needs to use extra chemical additives. The fundamentals of ultrasound, the advances of ultrasound in environmental remediation, the efficiency issues of ultrasound as well as the challenges and recent developments of ultrasound technology are summarized and discussed in more details in the next chapters of this book.

References

Chen D (2012) Applications of ultrasound in water and wastewater treatment. In: Chen D, Sharma SK, Mudhoo A (eds) Handbook on application of ultrasound: sonochemistry for sustainability. CRC Press, Taylor & Francis Group, Boca Raton

Estager J (2012) Integrating ultrasound with other green technologies: towards sustainable chemistry. In: Chen D, Sharma SK, Mudhoo A (eds) Handbook on application of ultrasound: sonochemistry for sustainability. CRC Press, Taylor & Francis Group, Boca Raton

Ghasemi Y, Rasoul-Amini S, Fotooh-Abadi E (2011) The biotransformation, biodegradation, and bioremediation of organic compounds by microalgae. J Phycol 47:969–980

Holland KS (2011) A framework for sustainable remediation. Environ Sci Technol 45:7116–7117

Khanal SK, Grewell D, Sung S, Van Leeuwen J (2007) Ultrasound applications in wastewater sludge pretreatment: a review. Crit Rev Environ Sci Technol 37:277–313

Megharaj M, Ramakrishnan B, Venkateswarlu K, Sethunathan N, Naidu R (2011) Bioremediation approaches for organic pollutants: a critical perspective. Environ Int 37:1362–1375

Pang YL, Abdullah AZ, Bhatia S (2011) Review on sonochemical methods in the presence of catalysts and chemical additives for treatment of organic pollutants in wastewater. Desalination 277:1–14

Pham TD, Shrestha RA, Virkutyte J, Sillanpää M (2009) Recent studies in environmental applications of ultrasound. Can J Civ Eng 36:1849–1858

Thangavadivel K, Megharaj M, Mudhoo A, Naidu R (2012) Degradation of organic pollutants using ultrasound. In: Chen D, Sharma SK, Mudhoo A (eds) Handbook on application of ultrasound: sonochemistry for sustainability. CRC Press, Taylor & Francis Group, Boca Raton

Chapter 2
Theory and Fundamentals of Ultrasound

Abstract The application of ultrasonic technology has been receiving wide attention by the world in wastewater treatment and environmental remediation areas. The use of ultrasound technology is shown to be very promising for the degradation of persistent organic compounds in wastewater as it is proven to be an effective method for degrading organic effluent into less toxic compounds. The advantages of this technology include potential chemical-free and simultaneous oxidation, thermolysis, shear degradation, enhanced mass-transfer processes together etc. Overall, sonochemical oxidation uses ultrasound to produce cavitation phenomena, which is defined as the phenomena of the formation, growth and subsequent collapse of microbubbles, releasing large magnitude of energy, and induces localized extreme conditions. The sonochemical destruction of pollutants in aqueous phase generally involves several reaction pathways such as pyrolysis inside the bubble and hydroxyl radical-mediated reactions at the bubble–liquid interface and/or in the liquid bulk. This chapter mainly reviews the fundamental of ultrasound technology.

Keywords Bulk region · Cavitation · Hot-spot theory · Interfacial region · Sonolysis · Ultrasonic waves

2.1 Theoretical Aspects of Ultrasound

During the past several years, ultrasound has been effectively applied as an emerging advanced oxidation process (AOP) for a wide variety of pollutants in wastewater treatment. It is proven to be an effective method for degrading organic effluents into less toxic compounds and able to mineralize the compounds completely in certain cases (Guo et al. 2010). The ultrasound process does not require addition of oxidants or catalyst, and does not generate additional waste streams as compared to adsorption or ozonation processes. Ultrasound process is also not affected by the toxicity and low biodegradability of compounds (Fu et al. 2007).

T. Y. Wu et al., *Advances in Ultrasound Technology for Environmental Remediation*,
SpringerBriefs in Green Chemistry for Sustainability, DOI: 10.1007/978-94-007-5533-8_2,
© The Author(s) 2013

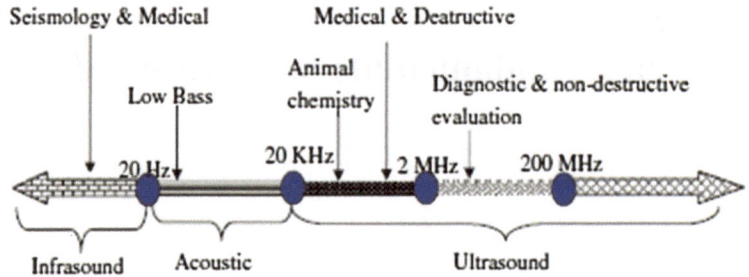

Fig. 2.1 Diagram of ultrasound range. Reprinted with permission from Pilli et al. (2011). Copyright (2011), Elsevier

Besides, ultrasonic degradation is claimed to be a non-random process, with cleavage taking place roughly at the center of the molecule and with degrading rate faster with larger molecule (Grönroos et al. 2008).

Ultrasonic waves (occurs at frequencies above 20 kHz) are a branch of sound waves and it exhibits all the characteristics properties of sound waves. Basically, they are classified into four different categories (namely, longitudinal/compressional waves, transverse/shear waves, surface/Rayleight waves, and plate/Lamb waves) based on the mode of vibration of the particle in the medium, with respect to the direction of the propagation of the initial waves (Raj et al. 2004). Depending on the frequency, ultrasound is divided into three categories, namely power ultrasound (20–100 kHz), high frequency ultrasound (100 kHz–1 MHz), and diagnostic ultrasound (1–500 MHz). Ultrasound ranging from 20 to 100 kHz is used in chemically important systems, in which chemical and physical changes are desired as it has the ability to cause cavitations of bubbles (Pilli et al. 2011; Rastogi 2011). Ultrasound ranging from 1 to 10 MHz is used for animal navigation and communication, detection of cracks or flaws in solids, and under water echo location, as well as diagnostic purposes (as shown in Fig. 2.1) (Pilli et al. 2011).

When applied on liquid, ultrasound waves consist of a cyclic succession of expansion (rarefaction) and compression phases imparted by mechanical vibration (Tang 2003). Compression cycles exert a positive pressure and push the liquid molecules together, while expansion cycles exert a negative pressure and pull the molecules apart (Vajnhandl and Marechal 2005). When pressure amplitude exceeds the tensile strength of liquid in the rarefaction regions, small vapor-filled voids called cavitation bubbles are formed (Chen 2012). Generally, pure liquids possess great tensile strengths and thus, available ultrasonic generators are unable to produce high enough negative pressures to cause cavitation. However, most of the liquids are usually impure and its tensile strength is reduced due to the presence of numerous small particles, pre-existing dissolved solids, and other contaminants. The impurities in liquid represent weak points in a liquid where nucleation of cavitation bubbles will occur (Vajnhandl and Marechal 2005). For instance, when pure water is used, more than 1,000 atm of negative pressure would

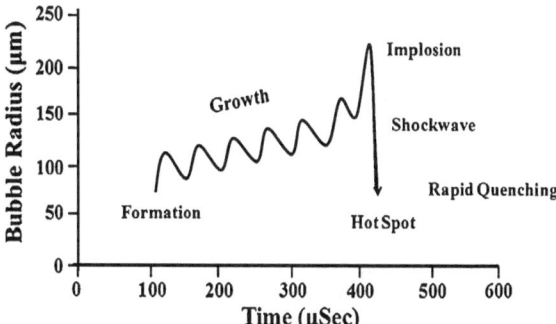

Fig. 2.2 Growth and implosion of cavitation bubbles in aqueous solution under ultrasonic irradiation. Reprinted with permission from Pang et al. (2011). Copyright (2011), Elsevier

be required for cavitation whereas for tap water, only a few atmosphere of pressure would be sufficient to form bubbles (Chowdhury and Viraraghavan 2009).

Once a bubble is created, two different cavitation phenomena which could take place in the liquid are: stable or transient cavitation. In stable cavitation, bubble wall couples with the acoustical field and oscillates about the equilibrium radius for several cycles. This occurs at low acoustic intensities, where the size of the bubble oscillates in phase with expansion and compression cycles and the bubbles grow slowly over many acoustical cycles (Thangavadivel et al. 2012). Due to its small variation in bubble size changes, this process is of little significance in terms of chemical effects (Destaillats et al. 2003). The process is also called rectified diffusion as during expansion, water vapor, dissolved gases and organic vapor will enter the bubble and will leave during contraction because of the effect of bubble surface area (Thangavadivel et al. 2012). When high intensity acoustic field is introduced, transient cavitation usually occurs. This causes growing cavitation bubble to eventually become unstable after a number of cycle and collapse during the compression cycle of ultrasonic wave. In this cavitation phenomena, the size of a bubble drastically increase from tens to hundreds of times the equilibrium radius before it collapses violently in less than a microsecond (Destaillats et al. 2003; Vajnhandl and Marechal 2005). Nevertheless, the classification of cavitation is vague as stable cavitation could lead to transient cavitation or transient cavitation could produce very small bubbles that undergo stable cavitation (Vajnhandl and Marechal 2005). In summary, phenomenon of cavitation consists of the repetition of three distinct steps: formation (nucleation), rapid growth (expansion) during the cycles until it reaches a critical size, and violent collapse in the liquid as shown in Fig. 2.2 (Pang et al. 2011).

The produced cavitation serves as a mean to concentrate the diffused sound energy. Either in low or high intensity acoustic field, once a cavity bubble experienced rapid growth and could no longer absorb the energy efficiently, the liquid will rush in and the cavity will eventually implode (Suslick 1989, 1990). Upon collapsing, each of the bubble would act as a hotspot, generating energy to increase the temperature and pressure up to 5,000 K and 500 atm, respectively, and cooling rate as fast as 10^9 K/s (Suslick 1990). The formation and growth of the cavitation bubbles is shown in Fig. 2.3. These collapsing bubbles create an unusual

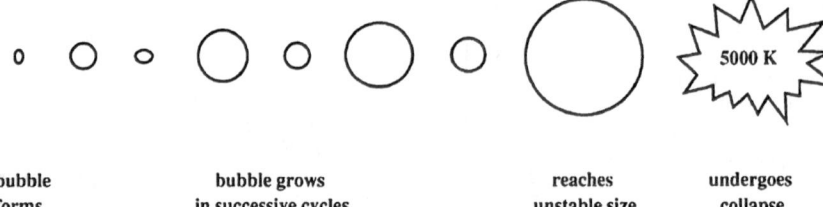

| bubble | bubble grows | reaches | undergoes |
| forms | in successive cycles | unstable size | collapse |

Fig. 2.3 Cavitation bubble formation, growth and collapse. Reprinted with permission from Chowdhury and Viraraghavan (2009). Copyright (2009), Elsevier

mechanism for high-energy chemical reactions due to enormous local temperatures and pressure (Suslick 1990).

There are many parameters which affect the cavitation and bubble collapse process and are listed as follows.

(a) Sound wave frequency: High frequency will reduce cavitational effect because (1) the negative pressure produced by rarefaction cycle is insufficient in duration and/or intensity to initiate cavitation or (2) compression cycle occurs faster than the time for microbubbles to collapse (Adewuyi 2001). At lower frequency, more violent cavitations will be produced, resulting in higher localized temperatures and pressure (Vajnhandl and Marechal 2005).

(b) Intensity of sound wave: Increasing intensity will increase the acoustic amplitude, resulting in a more violent cavitation bubble collapse (Adewuyi 2001).

(c) Solvent characteristics: Cavities are more readily formed in solvents with high vapor pressure, low viscosity, and low surface tension (Adewuyi 2001). However, the higher the vapor pressure, the less violent the bubbles collapse would be due to more vapor entering the bubbles (Peters 1996).

(d) Gas properties: Presence of soluble gases will result in the formation of larger number of cavitation nuclei. However, higher gas solubility would cause more gas molecules to diffuse into cavitational bubble, causing its collapse to be less violent (Vajnhandl and Marechal 2005). Heat capacity ratio (C_p/C_v) or polytropic ratio (γ) and thermal conductivity of the gas will also affect the amount of heat release during the collapse (Peters 1996; Adewuyi 2001).

(e) External pressure: Higher external pressure will reduce the vapor pressure of liquid and increases the intensity needed to induce cavitation (Vajnhandl and Marechal 2005).

(f) Temperature: For non-volatile substrates (that react through radical reaction in solution), reducing the reaction temperature will result in an increase in sonochemical reaction rates. The increase in cavitation intensity is caused by the lowering of vapor pressure and thus, reducing the amount of vapor diffusing into the bubbles to cushion the cavitational collapse (Destaillats et al. 2003; Adewuyi 2001).

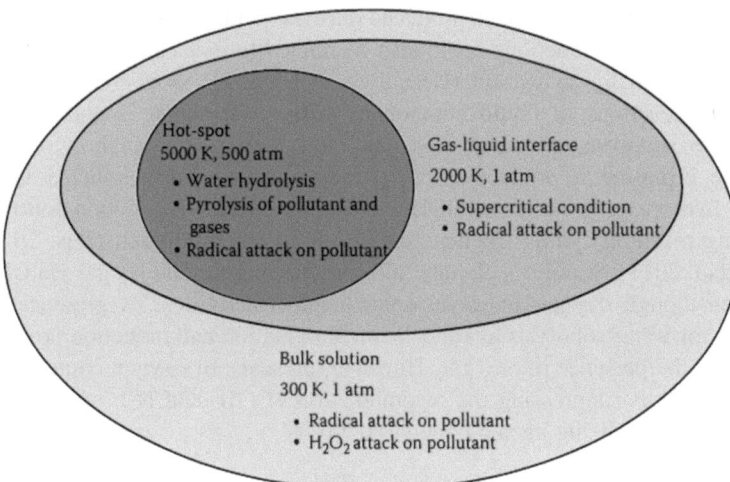

Fig. 2.4 Reaction zone in cavitation process. Reprinted with permission from Chowdhury and Viraraghavan (2009). Copyright (2009), Elsevier

There are four theories to explain sonochemical effects: (1) hot-spot theory; (2) "electrical" theory; (3) "plasma discharge" theory,and (4) supercritical theory. These theories lead to various mode of reactivity: pyrolytic decomposition, OH· oxidation, plasma chemistry, and supercritical water oxidation. Among these theories, hot-spot theory is widely accepted in explaining sonochemical reactions in the environmental field (Adewuyi 2001). According to hot-spot theory, each microbubble acts as a small microreactor which produces different reactive species and heat during its collapse (Vajnhandl and Marechal 2005). The temperature profile shows that there are three zones associated with a cavitational bubble (Chen 2012), as depicted in Fig. 2.4:

(a) Thermolytic center (hot spot), the core of the bubbles with localized hot temperature ($\sim 5,000$ K) and high pressure (~ 500 atm) during final collapse of cavitation. Inside this region, bubble water molecules are pyrolyzed forming OH· and H· in the gas phase. The substrate either reacts with the OH· or undergoes pyrolysis.
(b) Interfacial region between the cavitational bubble and bulk liquid, similar reaction as hot spot occurs, but in aqueous phase. However, additional reaction occurs in this region, in which OH· recombine to form H_2O_2. In this region, hydrophobic compounds are more concentrated than the bulk solution.
(c) The bulk region, the temperature remains at a level similar to room temperature because cavitation is an adiabatic process. In bulk phase, the reactions occurred are basically between the substrate and the OH· or H_2O_2

There is a clear distinction between the effects of ultrasound in homogeneous and heterogeneous media. For homogeneous media, sonochemical reactions are related to new chemical species produced during cavitation, whereas for the latter, enhancement of the reactions could also be related to mechanical effects induced in liquid system by sonication (Destaillats et al. 2003). Sonochemistry usually deals with reactions in liquid component. When ultrasound is applied, it will induce the sonolysis of water molecules and thermal dissociation of oxygen molecule, if present, to produce different kinds of reactive species such as OH·, H·, O· and hydroperoxyl radicals (OOH·). Reactive species production follows the following reactions, with ')))' denotes the ultrasonic irradiation (Eqs. 2.1–2.13) (Pang et al. 2011). Sonolysis of water also produces H_2O_2 and H_2 gas via OH· and H·. Even though oxygen improves sonochemical activities, its presence is not essential for water sonolysis as sonochemical oxidation and reduction process can proceed in the presence of any gas. However, presence of oxygen could scavenge the H· (and thus suppressing the recombination of OH· and H·), forming OOH·, which acts as oxidizing agents (Adewuyi 2001).

$$H_2O +))) \rightarrow OH \cdot + H \cdot \tag{2.1}$$

$$O_2 +))) \rightarrow 2O \cdot \tag{2.2}$$

$$OH \cdot + O \cdot \rightarrow OOH \cdot \tag{2.3}$$

$$O \cdot + H_2O \cdot \rightarrow 2OH \cdot \tag{2.4}$$

$$H \cdot + O_2 \rightarrow OOH \cdot \tag{2.5}$$

$$OH \cdot + H \cdot \rightarrow H_2O \cdot \tag{2.6}$$

$$2OH \cdot \rightarrow H_2O + O \cdot \tag{2.7}$$

$$OOH \cdot + OH \cdot \rightarrow O_2 + H_2O \tag{2.8}$$

$$2OH \cdot \rightarrow H_2O_2 \tag{2.9}$$

$$2OOH \cdot \rightarrow H_2O_2 + O_2 \tag{2.10}$$

$$H \cdot + H_2O_2 \rightarrow OH \cdot + H_2O \tag{2.11}$$

$$OH \cdot + H_2O_2 \rightarrow OOH \cdot + H_2O \tag{2.12}$$

$$2H \cdot \rightarrow H_2 \tag{2.13}$$

Besides chemical effects, ultrasound can also produce significant physical effects (sonophysical). When ultrasound is introduced, liquid medium will absorb the acoustic energy from sound waves and flow along the wave's propagation direction. Physical effects such as microstreaming, microstreamers, microjets, and shock waves can also be produced by cavitation bubbles, resulting turbulent fluid movement and a microscale velocity gradient in the vicinity of cavitational bubbles (Chen 2012).

(a) Microstreaming is the propagation of ultrasound waves through a liquid medium which creates small amplitude oscillatory motion of fluid elements around a mean position (Kuppa and Moholkar 2010). This phenomenon constitutes to an unusual type of fluid flow associated with velocity, temperature and pressure gradient (Tang 2003).

(b) Microstreamers are formed by cavitation bubbles travelling within the liquid to nodes or antinodes driven by *Bjerknes forces*. These bubbles travel in ribbon like structures along tortuous pathways (Chen 2012).

(c) Microjets are formed by the asymmetric collapse of cavitational bubbles near a micro-particle surface, with speed in the order of 100 m/s. The microjets will subsequently produce an asymmetric shock wave upon implosion of the bubble, resulting in direct erosion on particle's surface and de-aggregation of particles (Pang et al. 2011; Chen 2012).

(d) Shockwaves are produced by adiabatic compression of cavitational bubbles during the compression phase of radial motion. At the point of maximum compression, bubble wall comes to a sudden halt and rebounces at high velocity. The converging fluid elements are reflected back from bubble interface, creating a high pressure shock wave that propagates through the medium (Kuppa and Moholkar 2010).

The fluid movement produced by ultrasound could enhance the physical mass-transfer processes between solid-bulk and gas-bulk interfaces. Hence, these sonophysical effects described above can facilitate various mixing, breaking down of particles and macromolecules, polymer degradation, desorption, extraction, and cleaning processes (Chen 2012; Yasuda and Koda 2012).

2.2   Conclusion

Ultrasonic cavitation, which is an AOP, has been proposed as an attractive alternative method for the treatment of contaminants due to its advantages of being non-selective and without generating secondary pollutants. Four different theories are usually used to explain sonochemical effect but hot-spot theory is usually used to explain the process, in which microbubbles are produced to generate heat and different reactive species. Ultrasonic cavitation is known to generate reactive species such as OH·, H·, O· and OOH·, which are able to oxidize almost all toxic contaminants present in the environments. The mechanisms of ultrasound make it unique when compared with other AOPs. However, it is found that the degradation rate is rather slow by merely using ultrasonic treatment alone. Therefore, some efforts have been devoted to increase the degradation efficiency by applying hybrid techniques.

References

Adewuyi YG (2001) Sonochemistry: environmental science and engineering applications. Ind Eng Chem Res 40:4681–4715

Chen D (2012) Applications of ultrasound in water and wastewater treatment. In: Chen D, Sharma SK, Mudhoo A (eds) Handbook on application of ultrasound: sonochemistry for sustainability. CRC Press, Taylor & Francis Group, Boca Raton

Chowdhury P, Viraraghavan T (2009) Sonochemical degradation of chlorinated organic compounds, phenolic compounds and organic dyes—a review. Sci Total Environ 407:2474–2492

Destaillats H, Hoffmann MR, Wallace HC (2003) Sonochemical degradation of pollutants. In: Tarr MA (ed) Chemical degradation methods for wastes and pollutants. Environmental and industrial applications. Marcel Dekker, Inc., USA

Fu H, Suri RPS, Chimchirian RF, Helmig E, Constable R (2007) Ultrasound-induced destruction of low levels of estrogen hormones in aqueous solutions. Environ Sci Technol 41:5869–5874

Grönroos A, Pentti P, Hanna K (2008) Ultrasonic degradation of aqueous carboxylmethycellulose: effect of viscosity, molecular mass and concentration. Ultrason Sonochem 15:644–648

Guo W, Shi Y, Wang H, Yang H, Zhang G (2010) Intensification of sonochemical degradation of antibiotics levofloxacin using carbon tetrachloride. Ultrasound Sonochem 17:680–684

Kuppa R, Moholkar VS (2010) Physical features of ultrasound-enhanced heterogeneous permanganate oxidation. Ultrason Sonochem 17:123–131

Pang YL, Abdullah AZ, Bhatia S (2011) Review on sonochemical methods in the presence of catalysts and chemical additives for treatment of organic pollutants in wastewater. Desalination 277:1–14

Peters D (1996) Ultrasound in materials chemistry. J Mater Chem 6:1605–1618

Pilli S, Bhunia P, Yan S, LeBlanc RJ, Tyagi RD (2011) Ultrasonic pretreament of sludge: a review. Ultrason Sonochem 18:1–18

Raj B, Rajendran V, Palanichamy P (2004) Science and technology of ultrasonics. Pangbourne, UK

Rastogi NK (2011) Opportunities and challenges in application of ultrasound in food processing. Crit Rev Food Sci 51:705–722

Suslick KS (1989) The chemical effect of ultrasound. Sci Am 260:80–86

Suslick KS (1990) Sonochemistry. Science 247:1438–1445

Tang WZ (2003) Physicochemical treatment of hazardous wastes. CRC Press, US

Thangavadivel K, Megharaj M, Mudhoo A, Naidu R (2012) Degradation of organic pollutants using ultrasound. In: Chen D, Sharma SK, Mudhoo A (eds) Handbook on application of ultrasound: sonochemistry for sustainability. CRC Press, Taylor & Francis Group, Boca Raton

Vajnhandl S, Marechal AML (2005) Ultrasound in textile dyeing and the decolourization/mineralization of textile dyes. Dyes Pigments 65:89–101

Yasuda K, Koda S (2012) Development of sonochemical reactor. In: Chen D, Sharma SK, Mudhoo A (eds) Handbook on application of ultrasound: sonochemistry for sustainability. CRC Press, Taylor & Francis Group, Boca Raton

Chapter 3
Applications of Ultrasound Technology in Environmental Remediation

Abstract The use of ultrasound as one of the intensification technologies has undergone rapid development over the past decade. Among the many aspects in driving these developments, the increasing need to introduce environmentally friendly and clean technology, which is able to minimize contaminants at the source, is an important factor. Past studies show that ultrasound-assisted-chemical reactions have been carried out in many types of degradation reactions with high degradation rates and shorter reaction time as compared to conventional methods. Successful application of this technique to treat different types of halogenated hydrocarbons, pesticides, dyes, and other compounds has been widely reported in the literature. Many focus on addressing the drawbacks of onefold application of ultrasonic degradation by coupling with Fe^{2+}, H_2O_2, Fenton reagents, photocatalysts, and others. This chapter summarizes the results obtained from laboratory-scale studies, illustrating the promise and practicality of ultrasound as an effective advanced oxidation technique in solving environmental problems.

Keywords Carboxylic acids · Chlorinated compounds · Phenolic compounds · Pharmaceutical compounds · Dye wastewater · Pesticides · Polymer

3.1 Ultrasound Treatment of Chlorinated Compounds

Chlorinated solvents are the most-frequently found halogenated organic compounds in different types of water sources, with perchloroethylene (PCE) and its derivatives being among the most frequently detected in water wells. PCE has been commonly used as a cleaning solvent for most dry cleaners in the U.S. and has been reported as a major intermediate in the degradation of other chlorinated compounds (Sáez et al. 2011b). Sáez et al. (2011a) showed that high PCE

T. Y. Wu et al., *Advances in Ultrasound Technology for Environmental Remediation*, SpringerBriefs in Green Chemistry for Sustainability, DOI: 10.1007/978-94-007-5533-8_3, © The Author(s) 2013

degradation could be achieved using sonolysis operated at high frequencies (580 and 850 kHz). Besides major products of PCE degradation such as Cl^-, CO_2/CO, trichloroethylene (TCE), and dichloroethylene (DCE), new oxygenated by-products which have not been previously reported were identified, including trichloroacetic (TCAA) and dichloroacetic (DCAA) acids. Identification of these classes of products allowed a much more effective mass balance analysis. The authors proposed that the first step of the mechanism was the degradation of PCE into other chloroethenes in the hot bubbles' interior together with the production of OH·. This was then followed by two parallel steps, namely pyrolysis of the remaining compounds and oxidation by OH·. The latter predominated at low frequencies due to longer contact of chloroethenes and radicals inside the cavitating bubbles, leading to formation of highly soluble oxygenated compounds which were found to be chloroacetates. On the other hand, at high frequencies, the pyrolysis of chloroethenes inside the cavitation bubbles was prevalent. However, all of the compounds, especially water-soluble haloacetate ions, produced by sonolysis of PCE, showed some toxicity which was required to be completely removed by using a combination of different treatments (Sáez et al. 2011a). Thus, Sáez et al. (2010) carried out sonoelectrochemical degradation of PCE in aqueous sodium sulfate using controlled current density degradation sonoelectrolyses in batch mode. The synergic effect of the combined sonolysis and electrochemical treatment yielded 100 % fractional conversion of PCE and about 55 % degradation efficiency regardless of the ultrasonic intensity investigated (1.84–7.64 W/cm^2). Due to the significant reduction of treatment time, energetic consumption with sonoelectrochemical treatment was lower than the presented sonochemical or electrochemical treatments. Later, Sáez et al. (2011b) showed that significant enhancement of the efficiency of sonoelectrochemical process (100 % degradation) could be achieved in the absence of background electrolyte and at high ultrasonic power level. They concluded that this combined treatment should be able to provide economically viable treatment of PCE when high frequencies, pulsed ultrasound strategies, and/or flow sonoelectrochemical reactors were used (Sáez et al. 2010, 2011b).

Another chlorinated compound frequently detected in groundwater is trichloroethylene (TCE). TCE is used as an industrial solvent in metal degassing, paint stripping, and dry cleaning industries (Lee and Oh 2010). Lee and Oh (2010) studied the applicability of ultrasound to treat groundwater which was contaminated by volatile organic compound (VOC), with TCE and carbon tetrachloride (CT) chosen as target compounds. Degradation rate of TCE increased with decreasing initial concentration, while degradation rate of CT was constant regardless of the tested initial concentration. When sonolysis was performed on the aqueous mixture of TCE and CT, degradation rate of TCE was significantly enhanced while the degradation rate of CT was slightly inhibited. Enhancement of TCE degradation rate was due to the increase in OH· concentration by scavenging action of CT for H· and the generation of chlorine-containing radicals and HCIO from pyrolysis of CT by sonolysis (Lee and Oh 2010). Ayyildiz et al. (2007) showed that sonolysis degradation and mass transfer rates of TCE and ethylene bromide (EDB) increased with increasing power intensity. In general, higher power intensity not only enlarges cavitation bubbles but also generates higher temperature and pressure during bubble collapse. Their study also demonstrated that a theoretical mass transfer model of sonication system could be used to approximate the degradation rates of halogenated organic compounds by sonolysis (Ayyildiz et al. 2007). Rashid and Sato (2011) showed that photosonolysis of TCE, 1-trichloroethane, and tetrachloroethylene exhibited higher degradation efficiencies as compared to photolysis or sonolysis alone. Most of the compounds were degraded within the first 26 min and an increase in detection time to 60 min did not increase the VOC removal efficiencies. Statistical analysis revealed that the photosonolysis treatment efficiency was likely to be additive of photolysis and sonolysis treatment efficiency (Rashid and Sato 2011).

Katsumata et al. (2007) investigated the degradation of polychlorinated dibenzo-p-dioxins and polychlorinated dibenzofurans, namely 2,3,7,8-tetrachlorodibenzo-p-dioxins (TeCDD) by sonochemical treatment. A total of 93 % of TeCDD was decomposed in just 30 min using ultrasonic treatment (150 W and 20 kHz). Degradation pathway of TeCDD by ultrasonic would mainly consist of oxidative process by OH·. Under the presence of Fe(III) and UV irradiation, degradation rate of TeCDD increased to about four times (0.33 min^{-1}) than that obtained by ultrasound alone (0.08 min^{-1}) (Katsumata et al. 2007). Lim et al. (2011) investigated the effect of different frequency on sonolysis of chlorobenzene, chlorofoam, and carbon tetrachloride. For all the compounds, the highest degradation rate was recorded at 300 kHz. Higher degradation rate was found for compound with high volatility and in this case, carbon tetrachloride with the largest Henry's law constant exhibited the highest degradation as it was more easily diffused into the cavitation bubble in sonochemical process (Lim et al. 2011). For the degradation of chlorobenzene, Jiang et al. (2009) coupled reverberation ultrasound treatment with aerated biological treatment. The use of multiple frequencies (30, 60, and 100 kHz) in sonoreactor showed higher chlorobenzene degradation as compared to single frequency. This may be due to the formation of uniform sonicated field which decreased the dead regions inside the reactor and thus, improving the degradation efficiency (Jiang et al. 2009). Table 3.1 summarizes the performance of ultrasound-enhanced techniques for the degradation of various chlorinated compounds.

3.2 Ultrasound Treatment of Phenolic Compounds

Phenolic compounds, including the chloro- and nitro- derivatives and phenols are common organic contaminants which are released into the environment from effluents discharged by industries such as petroleum refining, coal gasification, pesticides, medications, pulp and paper, etc. They are also used as general disinfectants and as reagents in chemical analysis. Wastewater containing phenol, even in low concentrations, results in high levels of toxicity in the effluent stream

Table 3.1 Degradation performance of chlorinated compounds under ultrasonic-enhanced degradation in single/combined treatment

Chlorinated compound(s)	Treatment(s)	Initial concentration	Experimental conditions	Results	Reference
2,3,7,8-tetrachlorodibenzo-p-dioxins	Ultrasound	30 ng/L	20 kHz, 150 W, pH 2.8, 20 °C, 30 min	Degradation: ~87 %	Katsumata et al. (2007)
2,3,7,8-tetrachlorodibenzo-p-dioxins	Ultrasound + Fe + UV	30 ng/L	20 kHz, 150 W, pH 2.8, 20 °C, 30 min 2×10^{-4} M Fe(III)	Degradation: ~93 %	Katsumata et al. (2007)
Chlorobenzene	Ultrasound	16 mg/L	30 kHz, 100 W, 28 °C 8 h	Degradation: ~67.23 %	Jiang et al. (2009)
Chlorobenzene, chlorofoam, carbon tetrachloride	Ultrasound	0.05 mM	300 kHz, 18.8 W, 240 min	Degradation: ~65 % (Chlorobenzene); ~65 % (Chlorofoam); ~93 % (Carbon tetrachloride) TOC degradation: 35 % (Chlorobenzene); ~34 % (Chlorofoam); ~42 % (Carbon tetrachloride)	Lim et al. (2011)
Perchloroethylene	Ultrasound + electrochemical	~60 mg/L (362 µM)	20 kHz, 3.39 W/cm², 480 kWh/m³, 20 °C, 240 min 3.5 mA/cm², 0.12 W/ cm³, 7 kWh/m³	Fractional conversion: 100 % Degradation grade: 57 % Current efficiency: 14 % Cl mass balance error: 34 %	Sáez et al. (2011b)
Perchloroethylene	Ultrasound	~153.4 mg/ L (925 µM)	580 or 850 kHz, 1.436 W/cm², 0.061 W/cm³, 20 °C, 300 min	Fractional conversion: 99.02 % Degradation grade: 80.83 % Cl mass balance error: 10.1 %	Sáez et al. (2011a)

(continued)

Table 3.1 (continued)

Chlorinated compound(s)	Treatment(s)	Initial concentration	Experimental conditions	Results	Reference
Tetrachloroethane (TCA), trichloroethylene (TCE), tetrachloroethylene (TCE)	Ultrasound	100 mg/L	20 kHz, 193 W, 26 min	VOC removal: 12.7 % (TCA); 7.4 % (TCE); 6.8 % (PCE)	Rashid and Sato (2011)
Tetrachloroethane (TCA), trichloroethylene (TCE), tetrachloroethylene (TCE)	Ultrasound + UV	100 mg/L	20 kHz, 193 W, 26 min 35 W UV lamp, 1.65×10^{16} photons/s cm^3, 253.7 nm	VOC removal: 27.5 % (TCA); 55.6 % (TCE); 64.9 % (PCE)	Rashid and Sato (2011)
Trichloroacetic acid	Ultrasound + electrochemical	~ 817 mg/L (0.005 M)	24 kHz, 200 W, 3.8 W/cm^2, 600 min 0.024 W/cm^2	Fractional conversion: 97 % Degradation grade: 25.7 % Cl mass balance error: 16.3 %	Esclapez et al. (2010)
Trichloroethylene	Ultrasound	50 mg/L	20 kHz, 600 W, 35 W/cm^2, pH 5, 23–25 °C, 10 min	Degradation: ~40 %	Ayyildiz et al. (2007)
Trichloroethylene (TCE) and carbon tetrachloride (CT)	Ultrasound	10 mg/L	20 kHz, 54 W, 24 °C	Degradation rate constant: 0.0527 min^{-1} (TCE), 0.0482 min^{-1}	Lee and Oh (2010)

and gives foul odor to the water. Thus, it is classified as priority pollutant in the list of United States Environmental Protection Agency (USEPA) (Liu et al. 2009). Intensive attention has been paid to explore the degradation pathways to remove phenolic compounds from wastewater. Ultrasound has been investigated by many researchers as one of the techniques for phenol degradation. As shown in Table 3.2, ultrasound has been investigated by many researchers as one of the techniques in phenolic compound degradation.

Kubo et al. (2007) studied ultrasonic irradiation in the presence of composite particles of titanium dioxide (TiO_2) and activated carbon for phenol degradation and explained the effects of the amount of particles and the TiO_2 ratio in the particles on the degradation rate. They estimated the amount of absorbed phenol on the particles using the Langmuir adsorption isotherm and evaluated the degradation rate by the overall phenol concentration. They reported that a large amount of particles and high TiO_2 ratio in the composite particles could result in a high degradation rate.

The presence of zero valent metals (ZVMs) such as iron, copper, nickel, and zinc have played a significant role in enhancing the phenol degradation (Chand et al. 2009). A research was done to study the phenol degradation effectiveness with the presence of zero valent copper (metal pieces) and zero valent iron powder using 20, 300, and 520 kHz ultrasonic reactor. Complete phenol removal was observed within 25 min with the utilization of zero valent iron under 300 kHz, while with the usage of zero valent copper under 20, 300, and 520 kHz, the phenol removal was in the range between 10 and 98 % (Chand et al. 2009).

Coal ash could be used as a catalyst to generate OH· with the presence of H_2O_2/O_3, which enhances the phenol degradation. Liu et al.(2009) investigated the degradation of phenol under ultrasonic irradiation by coal ash and H_2O_2 or O_3. The combination of ultrasound/coal ash/H_2O_2 and ultrasound/coal ash/O_3 could achieve better performance for phenol degradation under more acidic and more alkaline conditions, respectively. Nakui et al. (2007) also investigated the effect of coal ash on degradation of phenol in water under ultrasonic irradiation conditions at 200 kHz and confirmed that coal ash (optimum amount at 0.4–0.6 wt %) could accelerate the degradation of phenol. Besides coal ash, ruthenium iodide (RuI_3) could also be utilized as a catalyst to study the influence of ultrasound on the oxidation of phenol due to its outstanding stability and high reusability (Rokhina et al. 2009).

The coupling of ultrasound with composite Fe-containing SBA-15 menostructured material (Fe_2O_3/SBA-15) and H_2O_2 is revealed to be a promising technique for phenol decay (Bremner et al. 2009). This combination is based on the catalyzed decomposition of H_2O_2 by an iron catalyst to form highly oxidizing OH·. Bremner et al. (2009) studied the benefits of combining high frequency with Fe_2O_3/SBA-15 in a Fenton-like system (sono-Fenton at high frequency) on the degradation of phenol solution by varying frequency values, ranging between 300 and 1,150 kHz. Composite Fe_2O_3/SBA-15 was also utilized by Segura et al. (2009) as heterogeneous catalyst to study the oxidation of phenolic aqueous solutions by coupling ultrasound with sono-Fenton and photo-assisted Fenton-like processes. A

Table 3.2 Degradation performance of phenolic compounds under ultrasonic-enhanced degradation in single/combined treatment

Phenolic compound	Treatment(s)	Initial concentration	Experimental conditions	Results	Reference
2,4-Dichlorophenoxyacetic acid	Ultrasound + stirring + catalyst (Fe) + H_2O_2	235 mg/L	20 kHz ultrasonic processor, 45 W, 300 mL cylindrical glass reactor, 22 °C, pH 3, 60 min 0.12 g powdered iron 1.7 mL H_2O_2	TOC removal: 60 %	Bremner et al. (2008)
2,4-Dinitrophenol	Ultrasound + O_3	20 mg/L	20 kHz ultrasonicator, 44 W, 75 mL glass vessel, 25 °C, pH 4, 250 min 1 mL/s O_3	Degradation efficiency: 100 % Degradation rate: 7.6×10^{-2} min^{-1}	Guo et al. (2008)
2,4,6-Trichlorophenol	Ultrasound	50 ppm (50 mg/L)	20 kHz direct-immersion horn sonicator, 50 °C	Degradation efficiency: 70 %	Joseph et al. (2011)
2,4,6-Trichlorophenol	Ultrasound + UV	50 ppm (50 mg/L)	20 kHz direct-immersion horn sonicator, 10 mW/cm^2, 50 °C 6 mW/cm^2 UV BLB lamp (365 nm)	Degradation efficiency: 76 %	Joseph et al. (2011)
2,4,6-Trichlorophenol	Ultrasound + Air	500 ppm (500 mg/L)	20 kHz ultrasonic horn, 180 W, 100 ml ultrasonic horn-type reactor, pH 4.8 0.0418 m^3/h air flow rate	Degradation efficiency: 65 % TOC reduction: 58.8 %	Shriwas and Gogate (2011a)

(continued)

Table 3.2 (continued)

Phenolic compound	Treatment(s)	Initial concentration	Experimental conditions	Results	Reference
2,4-Dichlorophenol	Ultrasound + O$_2$	50 mg/L	489 kHz, 20 W, 20 °C, pH 6.3	Degradation efficiency: 100 %, Degradation rate: 0.86 × 10^{-3} s^{-1}	Md. Uddin and Hayashi (2009)
2,4-Dichlorophenol	Ultrasound + catalyst (zero valent iron) + EDTA	100 mg/L	20 kHz sonicator, 385 W, 20 °C, 180 min 0.32 mM EDTA 50 g/L iron	Degradation efficiency: 100 % TOC removal: 81 %	Zhou et al. (2008)
2Chloro-5methyl phenol	Ultrasound + catalyst (TiO$_2$) + H$_2$O$_2$	5 mg/L	33 kHz ultrasonic bath, 1,255 W, 45 °C, pH 3, 120 min 60 mg/L TiO$_2$ 250 mg/L H$_2$O$_2$	Degradation efficiency: 98 % Degradation rate: 2.66 × 10^{-2} min^{-1} COD degradation: 70 %	Laxmi et al. (2010)
4-Cumylphenol	Ultrasound	5 mg/L	300 kHz piezoelectric disc, 80 W, cylindrical glass reactor, 20 °C, pH 6.5, 40 min Saturated argon gas	Degradation efficiency: 98 %	Chiha et al. (2011)
4-Chlorophenol	Ultrasound	5.18 mM (665.94 mg/L)	516 kHz ultrasonic reactor, 38.3 W, 21 °C, pH 5.5, 360 min	Degradation Efficiency: 63 %	Hamdaoui and Naffrechoux (2008)

(continued)

Table 3.2 (continued)

Phenolic compound	Treatment(s)	Initial concentration	Experimental conditions	Results	Reference
4-Chlorophenol	Ultrasound + UV	5.18 mM = 665.94 mg/L	516 kHz ultrasonic reactor, 38.3 W, 21 °C, pH 5.5, 360 min 15mW cm^{-2} low-pressure mercury lamp (254 nm)	Degradation Efficiency: 100 % Degradation rate constant: 9.89 × 10^{-2} min^{-1}	Hamdaoui and Naffrechoux (2008)
4-Chlorophenol	Ultrasound + H$_2$O$_2$ + catalyst (Cu/Al)	200 ppm = 200 mg/L	20 kHz sonicator, 100 W, 28 °C, pH 7.2 1,600 ppm H$_2$O$_2$, 1 g/L catalyst loading	4-CP removal: ~ 100 % TOC removal: 35 %	Kim et al. (2007)
4-Chlorophenol	Ultrasound + H$_2$O$_2$ + mill scale	100 mg/L	200 kHz sonicator, 200 W, 161 mm transducer diameter, 20 °C, pH 3.0, 60 min 100 mg/L H$_2$O$_2$, 1 g/L mill scale powder	Degradation Efficiency: 100 %	Liang et al. (2007)
4-Chlorophenol	Ultrasound + UV + catalyst	0.125 mM = 16.07 mg/L	20 kHz ultrasonic horn-type transducer, 51 W, 19 mm horn diameter, 25 °C, pH 5 100 mg/60 mL catalyst loading	Degradation Efficiency: ~75 % TOC removal: 35 %	Neppolian et al. (2011)

(continued)

Table 3.2 (continued)

Phenolic compound	Treatment(s)	Initial concentration	Experimental conditions	Results	Reference
4-Chlorophenol	Ultrasound + EDTA + catalyst (zero valent iron)	100 mg/L	20 kHz sonicator, 385 W, 20 °C, pH 7.4, 60 min 25 g/L zero valent iron 0.32 mM EDTA	Degradation efficiency: 100 % TOC removal: 73.3 % EDTA degradation efficiency: 81.4 %	Zhou et al. (2009)
Bisphenol A	Ultrasound	10 μM (2.28 mg/L)	300 kHz transducer, 0.190 W/ml, 150 ml glass reactor, pH 3, 60 min	Degradation efficiency: ~100 % Decay rate constant: – 9.14×10^{-2} min^{-1}	Gultekin and Ince (2008)
Bisphenol A	Ultrasound + O_3	100 g/L (0.1 mg/L)	20 kHz, 60 W/cm^2, 75 ml glass vessel, 25 °C, pH 6.5, 60 min 10 ml/min O_3 flowrate	Degradation efficiency: ~100 % Decay rate constant: 6.3×10^{-3} min^{-1}	Guo and Feng (2009)
Bisphenol A	Ultrasound	0.5 mM (114.15 mg/L)	404 kHz transducer, 12.9 kW/m^2, 25 °C, 10 h	Degradation efficiency: ~100 % TOC removal: ~20.2 %	Inoue et al. (2008)
Bisphenol A	Ultrasound	118 mol/L (26.94 mg/L)	300 kHz, 80 W, 300 ml cylindrical glass reactor, 20 °C, pH 3, 180 min Oxygen saturated	Degradation efficiency: ~100 % COD removal: 75 % TOC removal: 20 %	Torres et al. (2007a)

(continued)

Table 3.2 (continued)

Phenolic compound	Treatment(s)	Initial concentration	Experimental conditions	Results	Reference
Bisphenol A	Ultrasound + H_2O_2 + catalyst (FeSO$_4$)	118 mol/L (26.94 mg/L)	300 kHz, 80 W, 300 ml cylindrical glass reactor, 20 °C, pH 3, 180 min Oxygen saturated 119 mol/h H_2O_2, 100 µmol/L FeSO$_4$	Degradation efficiency: ~100 % COD removal: 40 % TOC removal: 5 %	Torres et al. (2007a)
Bisphenol A	Ultrasound	118 mol/L (26.94 mg/L)	300 kHz, 80 W, 20 °C, pH 3 Oxygen saturated	Degradation efficiency: 66 % COD removal: 35 % TOC removal: 10 %	Torres et al. (2007b)
Bisphenol A	Ultrasound + UV + catalyst (Fe(II))	118 mol/L (26.94 mg/L)	300 kHz, 80 W, 20 °C, pH 3 Low-pressure mercury lamp (254 nm) 100 µmol/L catalyst loading Oxygen saturated	Degradation efficiency: ~100 % COD removal: ~100 % TOC removal: ~100 %	Torres et al. (2007b)
Bisphenol A	Ultrasound	118 mol/L (26.94 mg/L)	300 kHz, 80 W, 500 ml cylindrical glass vessel, 20 °C, 500 min Oxygen saturated	Degradation efficiency: ~100 % COD removal: <50 % TOC removal: <20 %	Torres et al. (2008)
Bisphenol A	Ultrasound	118 mol/L (26.94 mg/L)	300 kHz piezo-electric disk, 80 W, 22 °C, pH 3, 120 min Oxygen saturated	Degradation efficiency: ~100 %	Torres-Palma et al. (2010)

(continued)

Table 3.2 (continued)

Phenolic compound	Treatment(s)	Initial concentration	Experimental conditions	Results	Reference
Bisphenol A	Ultrasound + UV + catalyst (Fe(II)) + adsorption (TiO$_2$) + H$_2$O$_2$	118 mol/L (26.94 mg/L)	300 kHz piezo-electric disk, 80 W, 22 °C, pH 3, 75 min. Solar lamp CPS suntest system. Oxygen saturated. 10 μmol/L L Fe. 50 mg/L TiO$_2$. 1.68 μmol/L min H$_2$O$_2$	Degradation efficiency: ~100 %	Torres-Palma et al. (2010)
Bisphenol A	Ultrasound	1 mg/L	800 kHz ultrasonator, 3 W/cm^2, 250 ml cylindrical glass vessel, 30 °C, pH 7, 60 min	Degradation efficiency: ~100 %. Decay rate constant: 6.87×10^{-2} min^{-1}	Zhang et al. (2011a)
Bisphenol A	Ultrasound + H$_2$O$_2$	1 mg/L	800 kHz ultrasonator, 3 W/cm^2, 250 ml cylindrical glass vessel, 30 °C, pH 7, 60 min. 0.1 mmol/L H$_2$O$_2$	Degradation efficiency: ~100 %. Decay rate constant: 9.08×10^{-2} min^{-1}	Zhang et al. (2011a)
p-aminophenol	Ultrasound + O$_3$	10 mmol/L (1,091.3 mg/L)	300 kHz ultrasonic processor, 0.3 W/L, glass reactor, 25 °C, pH 11, 720 min. 5.3 g/h O$_3$ dosage	Degradation efficiency: 99 %. TOC removal: 77 %. Degradation rate: 1.6×10^{-1} min^{-1}	He et al. (2007b)

(continued)

Table 3.2 (continued)

Phenolic compound	Treatment(s)	Initial concentration	Experimental conditions	Results	Reference
Phenol	Ultrasound + H_2O_2 + catalyst (Fe_2O_3/SBA-15)	2.5 mM (235.28 mg/L)	584 kHz ultrasonic processor, 0.029 W/mL, 25 °C, 360 min 0.6 g/L Fe_2O_3/SBA-15 1.19 g/L H_2O_2	TOC reduction: 29 %	Bremner et al. (2009)
Phenol	Ultrasound + H_2O_2 + Zero valent catalyst (iron)	2.5 mM (235.28 mg/L)	300 kHz ultrasound reactor, 12.9 W, 100 mL cylindrical glass reactor, pH 3, 20 °C, 60 min 2.38 g/L H_2O_2 0.6 g/L zero valent iron	Degradation efficiency: ~100 % TOC mineralisation: 45 %	Chand et al. (2009)
Phenol	Ultrasound + UV + H_2O_2 +catalyst (TiO_2)	10 g/L (10,000 mg/L)	25 kHz ultrasound reactor, pH 2, 180 min 8 W UV tube (365 nm) 2 g/L TiO_2 1 % H_2O_2	Conversion: 37.75 %	Khokhawala and Gogate (2010)
Phenol	Ultrasound + H_2O_2 + catalyst (CuO)	1 g/L (1,000 g/L)	25 kHz ultrasound bath, 1 kW, 90 min 1 g/L H_2O_2 1.5 g/L CuO	Degradation Efficiency: 42 %	Khokhawala and Gogate (2011)

(continued)

Table 3.2 (continued)

Phenolic compound	Treatment(s)	Initial concentration	Experimental conditions	Results	Reference
Phenol	Ultrasound + UV + O_3	2.5 mM (235.28 mg/L)	300 kHz ultrasonic processor, 0.86 W/cm^2, 150 ml reactor, pH 10, 90 min 108 kW Hg lamp (254 nm) 2 mg/L O_3	Decay rate coefficients: 8.69×10^{-2} min^{-1}	Kidak and Ince (2007)
Phenol	Ultrasound + adsorption (TiO_2 and activated carbon)	1 mol/m^3 (94.11 mg/L)	20 kHz ultrasonic processor, 75 W, 20 °C, 180 min Composite A (ratio activated carbon, amorphous TiO_2, anatase-type TiO_2 is 1:1:1) dose 90 g/dm^3	Degradation Efficiency: 100 %	Kubo et al. (2007)
Phenol	Ultrasound + coal ash + H_2O_2	10 mg/L	40 kHz ultrasound bath, 500 W, 22.5 L reactor, pH 6, 25 °C, 300 min 1.0 g/L coal ash 2.0 mM H_2O_2	Degradation efficiency: 83.4 %	Liu et al. (2009)
Phenol	Ultrasound + coal ash + O_3	10 mg/L	40 kHz ultrasound bath, 500 W, 22.5 L reactor, pH 6, 25 °C, 300 min 1.0 g/L coal ash 1.0 dm^3/min O_3	Degradation efficiency: 88.8 %	Liu et al. (2009)

(continued)

Table 3.2 (continued)

Phenolic compound	Treatment(s)	Initial concentration	Experimental conditions	Results	Reference
Phenol	Ultrasound	1 mg/L	130 kHz ultrasonic processor, 2.5 W/ cm^2, 2 L reactor, pH 3, 32 °C, 300 min	Decay rate coefficients: 0.0018 min^{-1} Degradation Efficiency: 100 %	Maleki et al. (2007)
Phenol	Ultrasound	10 mg/L	200 kHz ultrasonic processor, 200 W, 180 ml cylindrical reactor, 20 °C, 60 min	Degradation Efficiency: 70 %	Nakui et al. (2007)
Phenol	Ultrasound + coal ash	10 mg/L	200 kHz ultrasonic processor, 200 W, 180 ml cylindrical reactor, 20 °C, 60 min 0.5 wt % coal ash	Degradation Efficiency: 85 %	Nakui et al. (2007)
Phenol	Ultrasound + H$_2$O$_2$ + catalyst (Ruthenium iodide)	100 ppm (100 mg/L)	24 kHz sonicator, 360 min 1 g/L ruthenium iodide	Conversion: 70 %	Rokhina et al. (2009)
Phenol	Ultrasound + H$_2$O$_2$ + Catalyst (Fe$_2$O$_3$/SBA-15)	2.5 mM (235.28 mg/L)	20 kHz ultrasound sonicator, 0.13 W/ mL, pH 3, 22 °C, 180 min 35 mM H$_2$O$_2$ 0.6 g/L Fe$_2$O$_3$/SBA-15	Degradation efficiency: ~100 % TOC reduction: 30 %	Segura et al. (2009)

(continued)

Table 3.2 (continued)

Phenolic compound	Treatment(s)	Initial concentration	Experimental conditions	Results	Reference
Phenol	Ultrasound + UV + Catalyst (Fe$_2$O$_3$/SBA-15)	2.5 mM (235.28 mg/L)	20 kHz ultrasound sonicator, 0.13 W/mL, pH 3, 22 °C, 90 min 150 W mercury lamp 35 mM H$_2$O$_2$ 0.6 g/L Fe$_2$O$_3$/SBA-15	Degradation efficiency: ~100 % TOC reduction: 45 %	Segura et al. (2009)
Phenol	Ultrasound + Microwave + H$_2$O$_2$	1,250 ml of 1 mM (75.29 mg/L)	850 kHz ultrasonic processor, 2 W/cm^2, 2 L reactor, pH 3, 27 °C, 120 min Microwave 360 W, 93 °C 20 mmol H$_2$O$_2$	Degradation Efficiency: 76 %	Wu et al. (2008)
p-nitrophenol	Ultrasound + UV + catalyst (TiO$_2$) + H$_2$O$_2$	10 ppm (10 mg/L)	25 kHz ultrasonic bath, 195 W, stainless steel reactor, 30 °C, pH 2.5, 11 W UV tube, 0.5 g/L TiO$_2$, 1 g/L H$_2$O$_2$	Extent of degradation: 94.6 %	Mishra and Gogate (2011a)
p-nitrophenol	Ultrasound + catalyst (FeSO$_4$) + H$_2$O$_2$	0.5 % w/v (5 mg/L)	25 kHz ultrasonic bath, 1 kW, 28 °C, pH 3.7, 90 min 1 g/L FeSO$_4$ 5 g/L H$_2$O$_2$	Extent of degradation: 66.4 %	Pradhan and Gogate (2010)

total phenol degradation was achieved by applying ultrasound, followed by UV–visible light irradiation sequentially (Segura et al. 2009).

Kidak and Ince (2007) conducted an experiment to investigate the effect of ultrasonic cavitation at 300 kHz on the enhancement of ozone and UV-mediated decomposition of phenol by monitoring the concentration of phenol during 90-min exposure to ozonation, sonification, UV photolysis, O_3/ultrasound, UV/ultrasound, and O_3/UV/ultrasound operations. Ultrasound could also be combined with microwave irradiation to enhance the phenol degradation rate and efficiency (Wu et al. 2008). This combination was used to destroy phenol compound in aqueous solutions efficiently via sono-generated OH· and H_2O_2, in conjunction with the rapid thermal effect of microwaves on polar chemicals. Ultrasonic and UV irradiations combinations were also applied by Khokhawala and Gogate (2010) to investigate the degradation of phenol with the presence of TiO_2. The result obtained indicated that the combination favored in acidic condition and gave better phenol degradation as compared to the individual operation.

3.2.1 4-Chlorophenol

4-Chlorophenol (synonyms: *p*-chlorophenol, 4-hydroxychlorobenzene) or 4-CP is a harmful aromatic halide, which is difficult to be decomposed. This compound is released into the environment as a by-product of various industrial activities, including the chlorinated bleaching of paper and via waste from coal, gas, and oil industries. Due to its high toxicity, pretreatment is needed to remove 4-CP from wastewater before disposing into the environment.

A Fenton-like reaction system and ultrasonic irradiation was combined to investigate the degradation of 4-CP by utilizing three types of solid-Fe-containing catalysts, namely iron powder, basic oxygen furnace (BOF) slag, and mill scale (Liang et al. 2007). Remarkable results were obtained when iron or mill scale powder was added into 4-CP solution at pH 3 with complete degradation within 2 min of ultrasonic irradiation, while BOF slag showed no catalytic effect on 4-CP degradation (Liang et al. 2007). This phenomenon obtained is due to the higher concentration of calcium and lower concentration of iron found in BOF slag, which reduces the acidity of solution on dissolving. The decay of concentrated 4-CP solution by ultrasound only, UV irradiation only, and ultrasound/UV combination was investigated and the results obtained indicated that the degradation efficiency was higher by coupling ultrasound/UV process (100 %) as compared to a single operation with 63 and 67 % degradation of 4-CP degraded using ultrasound and UV irradiation, respectively (Hamdaoui and Naffrechoux 2008).

Bi_2O_3/$TiZrO_4$ was utilized as a catalyst to study the sonochemical and sonophotocatalytic degradation of 4-CP under visible light irradiation. The results showed a high efficiency for sonophotocatalytic degradation of 4-CP in the presence of visible light (Neppolian et al. 2011). Both sonochemical and photocatalysis processes were dependent on the solution pH while pH did not influence the rate of degradation under sonophotocatalytic process.

The beneficial role of using ultrasound in heterogeneous Fenton-like system for degradation of p-chlorophenol was investigated using three types of copper catalysts, namely CuO, Cu/Al_2O_3 (Cu/Al), and $CuO.ZnO/Al_2O_3$ (Cu/Zn) with an addition of H_2O_2 (Kim et al. 2007). Among the catalysts used, Cu/Al provided the most promising catalytic performance by showing the highest 4-CP degradation and TOC removal. Zhou et al. (2009) studied the competitive degradation relationship of 4-CP and EDTA in the ultrasound/Fenton-like system using iron supported heterogeneous catalyst or zero valent iron (ZVI). This research revealed weak competitive degradation relationship and both contaminants degraded rapidly at neutral circumstance without any addition of H_2O_2 because H_2O_2 was self-produced through oxygen activation catalyzed by iron/EDTA ligand reactions.

3.2.2 Bisphenol A

Bisphenol A (BPA) is a chemical compound largely used in the plastic industry as a monomer for the production of epoxy resins and polycarbonate, which contains BPA at a concentration of about 100 mg/L (Torres et al. 2007a; Inoue et al. 2008). However, low concentration of BPA may disturb the behavior of aquatic life if it is discharged into the environment without undergoing any treatment.

The possibility to achieve high degradation of BPA was studied intensively, so that the best conditions (such as ultrasonic frequency, power intensity, power density, initial BPA concentration, pH, etc.) could be determined (Gultekin and Ince 2008; Torres et al. 2008; Guo and Feng 2009; Zhang et al. 2011a). Degradation of BPA using sonochemical reaction was investigated by Inoue et al. (2008) by taking into consideration the relationship between BPA degradation and formation of H_2O_2 and nitric acid. Formic acid, propionic acid, and formaldehyde were obtained as products from BPA degradation.

BPA degradation was studied by Torres et al. (2007b) under the combination of ultrasound with Fe^{2+}, ultrasound/UV and ultrasound/UV/Fe^{2+}. This research showed that the difficulties in obtaining mineralization of BPA through ultrasound could only be overcome by the combination of ultrasound/UV/Fe^{2+}. Iron (II), Fe^{2+} was also utilized to explore the degradation of BPA, by advanced oxidation process that combined ultrasound, photo-Fenton, and TiO_2 photocatalysis system (Torres-Palma et al. 2010). The combination of these three operations resulted in an efficient and fast mineralization of BPA.

BPA was also treated by Guo and Feng (2009) using ultrasound and O_3 with different flow rates. This combination enabled full removal of BPA after 60 min of treatment, whereas only 34.6 and 63 % of BPA were degraded in the single ultrasound and O_3 system, respectively. The effect of different gases (oxygen, air, and argon) on the initial BPA degradation was investigated by Torres et al. (2008) at 300 kHz and 80 W. The results obtained indicated that oxygen as saturating gas gave the highest effect on degradation rate, followed by air and argon. Another study was carried out by Zhang et al. (2011a) to investigate the degradation of BPA by ultrasonic irradiation in the presence of different additives, namely H_2O_2, air bubbles, and humic acid. The addition of low concentration of H_2O_2 could facilitate BPA degradation efficiently, while the presence of humic acids and aeration could inhibit the degradation rate.

3.2.3 Other Phenolic Compounds

3.2.3.1 2,4,6-Trichlorophenol

2,4,6-Trichlorophenol (synonyms: TCP, phenaclor) is a chlorinated phenol that has been used widely in the leather industry, wood preservatives, glue preparations, and also as an intermediate in the preparation of pesticides. Most uses of TCP were discontinued due to its toxicity but several fungicides still require the use of TCP in synthesis (Joseph et al. 2011). TCP must be decomposed before being discharged into environments. Shriwas and Gogate (2011a) investigated the degradation of TCP using two types of sonochemical reactors, namely ultrasonic horn and ultrasonic bath reactors, by applying different operating parameters (temperature, power input, and pH) and additives (solid particles, air, and H_2O_2). Maximum degradation was observed in the presence of air in the horn-type reactor. Operating parameters and synergistic effects on sonolytic, photolytic, and the combination of ultrasound and UV irradiation (sonophotolytic) in the degradation of TCP were studied by Joseph et al. (2011). They reported that the degradation of TCP was proportional to an increase in acoustic intensity and UV intensity. TCP concentration was reduced after sonophotolysis treatment, as compared to a single operation of photolysis or sonolysis (Joseph et al. 2011).

3.2.3.2 2,4-Dichlorophenol

2,4-Dichlorophenol (2,4-DCP) is primarily used as an intermediate in the preparation of herbicide 2,4-dichlorophenoxyacetic acid (2,4-D), which exhibits high toxicity but low biodegradability (Uddin and Hayashi 2009). Zhou et al. (2008) reported the performance of the combination of ultrasound/Fe/EDTA system on DCP degradation by varying different initial conditions (such as DCP concentration, iron, and EDTA dosage as well as reaction temperature). The degradation rate constant of DCP using ultrasound/Fe/EDTA was 7 and 32 times higher than in Fe/EDTA and ultrasound system, respectively. Md. Uddin and Hayashi (2009) investigated the effects of dissolved gases and pH on the sonolysis of DCP. The presence of oxygen and argon enhanced the degradation rate significantly but the degradation rate could be inhibited by nitrogen gases.

3.2.3.3 p-Nitrophenol

p-Nitrophenol (p-Np) (synonyms: 4-nitrophenol, 4-hydroxynitrobenzene) is one of the most refractory contaminants present in industrial wastewater due to its high stability and solubility in water. This substance is produced from industries such as textiles, pulp and paper, plastics, etc. (Mishra and Gogate 2011a). Pradhan and Gogate (2010) assessed the p-Np degradation under various operating parameters based on ultrasound, Fenton process, ultrasound/H_2O_2, ultrasound/Fe, ultrasound/FeSO$_4$, ultrasound/conventional Fenton process, and ultrasound/advanced Fenton process. Maximum degradation of p-Np was obtained at approximately 66.4 %, when 1 g/L FeSO$_4$ and 5 g/L H_2O_2 were used. The p-Np was also found to be degraded through the use of a combination of ultrasonic irradiation, ultraviolet radiation, and a semiconductor photocatalyst, which enhanced the formation of free radicals (Mishra and Gogate 2011a). In all the systems investigated, the maximum degradation of p-Np (94.6 %) happened at 10 ppm p-Np by using the combination of sonophotocatalysis with an addition of 1 g/L H_2O_2.

3.2.3.4 2,4-Dichlorophenoxyacetic acid (2,4-D)

2,4-Dichlorophenoxyacetic acid (2,4-D) has been widely used to control the broad-leaved weeds due to its low cost, high effectiveness, and moderate toxicity (Quan and Chen 2011). However, the solubility and non-volatility of 2,4-D are harmful to the environment and human health. Hence, this compound has to be controlled and treated before disposal. Bremner et al. (2008) evaluated the degradation performance of 2,4-D using acoustic or hydrodynamic cavitation in conjunction with the advanced Fenton process. Utilization of zero-valent iron and H_2O_2 were very effective in degrading high concentrations of 2,4-D. The addition of iron particles enhanced the cavitational intensity because the solid particles were acting as nuclei for surface cavitation, thereby increasing the number of cavitational events occurring in the reactor. However, an appreciable increase was observed in the presence of H_2O_2, which acts as a source for OH· by Fenton chemistry as well as by dissociation in the presence of ultrasound (Bremner et al. 2008).

3.2.3.5 2Chloro-5methyl phenol (2C-5MP)

2Chloro-5methyl phenol (2C-5MP) is widely used in manufacturing of resins, herbicides, pharmaceuticals, and tricresylic acid surfactants. It has high toxicity even in low concentrations. The degradation of this compound was investigated by Laxmi et al. (2010) in aqueous solution by using ultrasonification in the presence of TiO_2 and H_2O_2. Maximum degradation rate was achieved by ultrasonification/TiO_2/H_2O_2 combination as compared to ultrasonification/TiO_2 and ultrasonification/H_2O_2. Laxmi et al. (2010) also concluded that 2C-5MP is hydrophilic and thus reacts with OH· in solution, rather than inside because the bubble/liquid interface is hydrophobic.

3.2.3.6 2,4-Dinitrophenol

The USEPA recommends restricting 2,4-dinitrophenol (DNP) concentration in natural water below 10 ng/L due to its high toxicity and stability (Guo et al. 2008). The degradation mechanisms and reaction kinetics of DNP in different processes were proposed by Guo et al. (2008). Additives such as CuO, CCl$_4$, O$_3$, NaCl, and KI were observed to enhance the DNP degradation, while the presence of Na$_2$CO$_3$ inhibited the degradation process. This studies also reported that DNP was completely removed in ultrasound/O$_3$ after 1 h of treatment but only 4 and 77 % of DNP was removed in single treatment process of ultrasound and O$_3$, respectively.

3.2.3.7 4-Cumylphenol

4-cumylphenol (4-CyP) is 12 times more estrogenically active than BPA and is widely used as material for producing polycarbonate plastics, surfactants, fungicides, and preservatives (Chiha et al. 2011). The influence of operating parameters on ultrasound treatment of 4-CyP was investigated by Chiha et al. (2011). The extent of degradation was observed to be inversely proportional to the initial concentration of 4-CyP but the degradation rate increased proportionally with increasing temperature and ultrasonic power. The effect of saturating gas was also examined and 4-CyP degradation was enhanced in the presence of saturating gas, following the order: argon > air > nitrogen. The presence of bromide anions could also promote the effect of 4-CyP degradation (Chiha et al. 2011).

3.2.3.8 p-Aminophenol

p-Aminophenol (PAP) is used as an intermediate in the production of medicines, azo, sulfur, acid wood, and as photograph developer. It is identified as a serious environmental pollutant. PAP degradation by using ultrasound, ozonation, and a combination of both methods were examined by optimizing the operation conditions (He et al. 2007a). The highest degradation rate was obtained in the combined system than in the single treatment. However, the degradation efficiency decreased drastically when n-butanol was added to the combined system. Also, 4-iminocyclohexa-2,5-dien-1-one, phenol, but-2-enedioic acid and acetic acid were detected as intermediate products but only but-2-enedioic acid and acetic acid remained after undergoing ultrasonic oxidation up to 120 min (He et al. 2007a).

Table 3.3 Degradation performance of carboxylic acids under ultrasonic-enhanced degradation in single/combined treatment

Carboxylic acid	Treatment(s)	Concentration	Experimental conditions	Results	Reference
Acetic acid	Ultrasound	300 mg/L	40 kHz, 84 W, 43 °C, 60 min	Degradation: $\sim$9 %	Fındık and Gündüz (2007)
EDTA	Ultrasound + O_3	0.25 mM (73.06 mg/L)	20 kHz, 88 W, 150 mL, 20 °C, 75 min 0.5 g/h	Degradation: 70 %	Wang et al. (2010a)
EDTA	Ultrasound + Fe^0	1 mM (292.24 mg/L)	20 kHz, 440 W, pH 6.9–7.4, 20 °C, 60 min 25 g/L Fe^0, 1 L/min purified air	Degradation: 98 %	Zhou et al. (2010)
EDTA-Cu	Ultrasound + electrodeposition	110 mg/L (each)	20 kHz, 300 W, pH 7, 220 min 1 V/cm	Cu removal: 95.6 % COD removal: 84 %	Chang et al. (2009)
Formic acid	Ultrasound	3 M (138.09 mg/L)	607 kHz, 125 W, 0.26 W/mL, 90 mL/min Ar gas, 20 °C, 3 h	CO_2 yield: 0.65 μmol/kJ CO yield: 1.1 μmol/kJ H_2 yield: 0.35 μmol/kJ	Navarro et al. (2011)
Oxalate	Ultrasound	0.9 mM	358 kHz, 100 W, 0.6 L, pH 3, 15 °C, 400 min	Oxalate degradation: $\sim$30 % TOC degradation: $\sim$30 % Oxalate degradation rate: 8.3×10^{-7} M/min	Vecitis et al. (2010)
Oxalate	Ultrasound + O3	0.9 mM	358 kHz, 100 W, 0.6 L, pH 3, 15 °C, <60 min 350 μM O_3	Oxalate degradation: $\sim$100 % TOC degradation: $\sim$100 % Oxalate degradation rate: 2.8×10^{-5} M/min	Vecitis et al. (2010)
Salicylic acid	Ultrasound + TiO_2	10 μM (1.38 mg/L)	36 kHz, 200 W, 2 mL, 20 °C, 20 min 2 g TiO_2 (Rutile)	Final DHBA (degradation product) concentration: $\sim$1.7 μM	Shimizu et al. (2008)

EDTA Ethylenediaminetetraacetic acid, *DHBA* Dihydroxybenzoic acid

3.3 Ultrasound Treatment of Carboxylic Acids

Table 3.3 summarizes the performance of ultrasound-enhanced techniques for the degradation of various carboxylic acids. One of the examples of carboxylic acid often subjected to ultrasonic studies is ethylenediaminetetraacetic acid (EDTA). Wastewater containing EDTA has become an environmental concern as it is commonly used in pharmaceutical and agricultural industries. EDTA can potentially mobilize toxic heavy metals, extend its biological availability to aquatic life, and increase the risk posed by this metal to the water sources (Wang et al. 2010a). Effective treatment of EDTA is necessary as considerable amount of metal-EDTA complexes often pass through wastewater treatment facilities without efficient degradation (Zhou et al. 2010). Wang et al. (2010a) observed a remarkable synergistic effect when ozonolysis was combined with sonolysis for EDTA degradation. This effect was due to the increase of O_3 dissolution rate caused by the enhancement of gas–liquid mass transfer by ultrasound. Besides water and oxygen splitting, ultrasound induced ozone decomposition to provide an additional source of OH· for EDTA degradation. Main oxidation intermediates of EDTA identified was amino acids, such as amido acetic acid, N-methylamino acetic acid, N-ethylamino acitic acid, N-(2-aminoethyl)amino acetic acid, and others (Wang et al. 2010a). Zhou et al. (2010) investigated the role of ultrasound on zero-valent iron/air system for EDTA degradation. Excellent synergistic effect was also observed with calculated synergy factor of 7.8. Ultrasound enhanced the overall degradation rate by accelerating zero-valent iron corrosion and production of Fe^{2+}, enhancing H_2O_2 production through overcoming the kinetic barrier of oxygen activation by iron-EDTA complex and improving EDTA mineralization, while reducing harmful by-products. They postulated that the dominant oxidant generated in the combined system was ferryl-EDTA complex ($[Fe^{IV}O]EDTA$) rather than OH·, which was able to oxidize EDTA even under circumneutral conditions (Zhou et al. 2010). Ultrasound was also found to enhance electrodeposition treatment of EDTA-copper wastewater (Chang et al. 2009). The presence of ultrasound decreased the diffusion layer at the electrode surface, which effectively increased the reaction rate. According to Chang et al. (2009), the vibration caused by ultrasound also increased the collision frequency of molecular interaction and deposition rate but decreased the over-potential of the electrode (Chang et al. 2009).

Shimizu et al. (2008) investigated the generation of OH· during the ultrasonic irradiation of salicylic acid with the presence of TiO_2. Significant increase of salicylic acid derivatives, namely 2,3-dihydroxybenzoic acid (DHBA) and 2,5-DHBA, were recorded when ultrasonic irradiation was applied in the presence of TiO_2. Addition of OH· scavengers such as dimethylsulfoxide (DMSO), methanol, and mannitol could suppress the production of DHBA, with DMSO showing the best suppressive effect. When the reaction vessel was degassed by the application of 96 kPa negative pressure, the generation of DHBA was almost negligible, indicating that the production of OH· was completely suppressed. This finding showed that the presence of TiO_2 was able to promote the generation of OH·

during ultrasonic irradiation and the process could be mediated through the induction of cavitation bubbles in irradiating solutions (Shimizu et al. 2008).

Navarro et al. (2011) presented a study on the effect of ultrasonic frequency on formic acid sonochemical degradation. They found that not only CO_2, CO, H_2, and oxalic acid were produced, but formaldehyde and methane were also detected. An increase of approximately six- to eightfolds of the total formic acid degradation yield was recorded when the frequency was increased from 20 to 607 kHz. This is because high frequency gave a more diffuse and widely distributed zone of cavitation with formation of larger cavitation bubbles. They concluded that one of the main differences between formic acid sonolysis at low and high frequencies was that the latter initiated Fischer–Tropsch hydrogenation of CO, leading to production of methane and formaldehyde as by-products (Navarro et al. 2011). Fındık and Gündüz (2007) investigated ultrasound degradation of one of the most resistant carboxylic acids to oxidation, namely acetic acid. An addition of 1.5 M NaCl provided a 60 % increase in degradation as compared to the treatment without adding NaCl. However, sonolysis degradation of acetic acid was still low (~ 9 %) as compared to the efficiency obtained by wet air oxidation (~ 100 %) (Fındık and Gündüz 2007).

Vecitis et al. (2010) showed that the combination of ozonolysis and ultrasound could degrade aqueous oxalate efficiently. Oxalate is commonly detected in terrestrial and aquatic environment. This compound is moderately recalcitrant toward oxidation and often accumulates in natural waters, leading to microbial growth. Using the combined process, synergistic effect was observed where the apparent oxalate oxidation rates were 16 times that of a simple linear addition of the two independent reaction systems. OH· was the only oxy-radical capable to oxidize oxalate in this study, in which case plausible OH· production mechanisms were evaluated by the authors to explain the synergism of a combination of ultrasound and oxonolysis toward bioxalate decomposition. A free-radical chain mechanism was proposed, whereby $HC_2O_4^-$ + OH· reaction acts as primary propagation step, while the termination step occurred through the O_3 + $CO_2 \cdot^-$ reaction via an O-atom transfer mechanism. Besides, no increase in degradation rates was observed with an addition of H_2O_2 during the course of the process (Vecitis et al. 2010).

3.4 Ultrasound Treatment of Polymer

Recently, novel degradation technology based on ultrasonication, which is less energy intensive and contaminative as compared to the thermal/peroxy induced degradation technology, has been applied to polymer solution (Guo and Peng 2007). According to Desai et al. (2008), the term "degradation" means breaking down of chemical structure in classical chemical usage, but in terms of polymer chemistry, this word seems to imply a decrease in molecular weight or intrinsic viscosity of the polymer solution. The use of ultrasonication for depolymerization of molecules is highly advantageous because it reduces the molecular weight

Table 3.4 Degradation performance of polymers under ultrasonic-enhanced degradation in single/combined treatment

Polymer compound(s)	Treatment(s)	Concentration	Experimental conditions	Results	Reference
Dextran	Ultrasound	$M_w = 1.51 \times 10^5$, $M_w/M_n = 3.0$	500 kHz, 22 W, 50 mL, 25 °C, 120 min	M_n reduction: ~98.2 %. Degradation rate constant: 28x10^7 mol/g.min	Koda et al. (2011)
Methyl cellulose	Ultrasound	$M_w = 1.98 \times 10^5$, $M_w/M_n = 2.8$	500 kHz, 22 W, 50 mL, 25 °C, 120 min	M_n reduction: ~98.6 %. Degradation rate constant: 50×10^7 mol/g.min	Koda et al. (2011)
Poly(butyl methacrylate)	Ultrasound	2 g/100 mL toluene, $M_w = 3.1 \times 10^5$, $M_n = 4.26$	20 kHz, 200 W, 20 °C, 200 min	M_w reduction: ~78.4 %. M_w/M_n reduction: ~53 %	Kanwal et al. (2007)
Poly(ethylene oxide)	Ultrasound	$M_w = 1.58 \times 10^5$, $M_w/M_n = 3.8$	500 kHz, 22 W, 50 mL, 25 °C, 120 min	M_n reduction: ~97.6 %. Degradation rate constant: 27×10^7 mol/g.min	Koda et al. (2011)
Poly(ethylene oxide) (PEO), poly(acrylic acid) (PAA), poly(vinyl pyrrolidone) (PVP)	Ultrasound + TiO$_2$	1 g/L (PEO,PAA), 5 g/L (PVP)	25 kHz, 180 W, 36 W/cm^2, 5 mL, 4 h 1 g/L TiO$_2$	Degradation rate constant: 2.58×10^{-7} mol/g.min (PEO), 1.48×10^{-7} mol/g.min (PAA), 4.24×10^{-7} mol/g.min (PVP)	Aarthi et al. (2007)
Poly(methyl methacrylate)	Ultrasound + UV	3.1×10^{-6} mol/L toluene	25 kHz, 1.2 W/cm^2, 480 min 125 W high pressure mercury vapor lamp, 365 nm, 10.5 μ Einstein/s, 965 μW/cm^2, 2.5 g/L benzoin	M_n reduction: ~53 %. M_w/M_n reduction: ~37 %	Vinu and Madras (2011)
Poly(methyl methacrylate), poly(ethyl methacrylate), poly(butyl methacrylate)	Ultrasound	2 g/50 mL toluene	25 kHz, 36 W/cm^2, 30 °C, 200 min	Degradation rate constant: 4.8×10^{-8} mol/g.min (Polymethyl methacrylate); 5.3×10^{-8} mol/g.min (Polyethyl methacrylate); 6.3×10^{-8} mol/g.min (Polybutyl methacrylate)	Daraboina and Madras (2009)

(continued)

Table 3.4 (continued)

Polymer compound(s)	Treatment(s)	Concentration	Experimental conditions	Results	Reference
Poly(styrene sulfonate)	Ultrasound	3.88×10^{-5} mol/L ethanol, $M_n = \sim 6 \times 10^4$	439 kHz, 6 W, 60 min	M_n reduction: ~92 %	Zhu et al. (2011)
Poly(vinyl pyrrolidone)	Ultrasound	$M_w = 1.3 \times 10^6$	35 kHz, 80 W, ~5.56 h	M_w reduction: ~95 %	Akyüz et al. (2008)
Poly(vinyl pyrrolidone)	Ultrasound	5 g/L mixed acetone and water, $M_w = 1.3 \times 10^5$, $M_n = \sim 6.7 \times 10^5$	24 kHz, 350 W, 25 °C, 100 min	M_n reduction: ~77.6 %	Mehrdad (2011)
Poly(γ-benzyl-L-glutamate)	Ultrasound	30.6 mg/60 mL DMAc/0.5 % LiCl	47 kHz, 185 W, 21 °C, 600 min	M_n reduction: ~42 %; M_w reduction: ~52 %; M_w/M_n reduction: ~17.3 %	Ostlund et al. (2008)
Polyethylene	Ultrasound	1 % w/v in 100 mL o-dichlorobenzene, Intrinsic viscosity ~54 mL/g	22.5 kHz, 240 W, 50 mL, 360 min	Intrinsic viscosity reduction: ~37 %	Desai et al. (2008)
Polyolefin elastomer melt (Propylene–ethylene random copolymers (PEE) and ethylene–octane random copolymers (POE))	Ultrasound	Melt flow rate = 2 g/10 min (PEE); 1 g/10 min (POE) Comonomer content = 15 wt % (PEE); ~20wt % (POE)	20 kHz, 200 W, 160 °C, 600 min	Intrinsic viscosity reduction: ~29 % (PEE); ~32 % (POE)	Chen et al. (2007a)
Polypropylene	Ultrasound	$M_w = 6.7 \times 10^5$, $M_w/M_n = 3.8$	25 kHz, 0.2 kW, 3 min	M_w reduction: ~47.7 %; M_w/M_n increment: ~18.4 %	Guo and Peng (2007)
Pullulan	Ultrasound	$M_w = 2.28 \times 10^5$, $M_w/M_n = 2.0$	500 kHz, 22 W, 50 mL, 25 °C, 120 min	M_n reduction: ~94 %; Degradation rate constant: 21×10^7 mol/g min	Koda et al. (2011)

M_w Average molecular weight, M_n Number average molecular weight, M_w/M_n Molecular weight distribution index, DMAc N,N-dimethyl acetamide

simply by splitting the most susceptible chemical bond without making any changes in the chemical nature of the polymer (Desai et al. 2008). When an ultrasonic wave passes through the polymer solution, both cyclic tensions and compressions, which causes cavitation, occur. Microbubbles formed by cavitation will collapse and produce intense sheer and shock waves on polymer molecules near the bubble. The polymer molecule near the collapsing microbubble will experience high shear force and move faster than the polymer molecule far from the cavitation. This relative motion of polymer chain and mechanical stress generated are responsible for the degradation of polymer (Desai et al. 2008, Daraboina and Madras 2009). Molecules longer than a critical length are subjected to this scission, while shorter molecules with molecular weight less than a critical value are resistant to the effects of ultrasonic irradiation. Hence, molecular weight of polymer decreases continuously upon irradiation until it reaches a limiting value, which depends on the sonication conditions (Akyüz et al. 2008; Daraboina and Madras 2009). Recent studies in using ultrasound technology for the degradation of various polymers are summarized in Table 3.4.

Poly(akyl methacrylate) is often subjected to ultrasonic studies due to its industrial versatility. For example, poly(methyl methacrylate) (PMMA) is a widely used thermoplastic in glass replacement, intraocular lenses, denture fixing, paints, and lubricating fluids. Thus, mechano-chemical degradation of these polymers can be performed to modify the properties of the end product and for the effective remediation of the waste plastic (Vinu and Madras 2011). For the degradation of two polymer samples, the lower molar mass PMMA and the higher molar mass poly(butyl methacrylate) (PBMA), Kanwal et al. (2007) presented a study on determining the existence and range of lower limiting molar mass degradation for chemically different polymers. They found that the value of the limiting molar mass of polymer was dependent on the frequency of the ultrasonic used and to some extent on the amount of entanglements present in the molecule chain but not on the chemical nature of the polymer. Thus, there was a limiting molar mass that ultrasonic degradation was not possible at the applied frequency of 20 kHz (Kanwal et al. 2007). A study on different poly(alkyl methacrylates) was also conducted by Daraboina and Madras (2009). They showed that the rate coefficient increased with increasing number of atoms in the alkyl group and in their experiment, the order of degradation was PBMA > poly(ethyl methacrylate) (PEMA) > PMMA. During ultrasonication, the movement of main chain in the polymer increases with the length of side chain, leading to molecular chain scission in the main chain of the polymer. During the ultrasonication of PBMA, an addition of different initiators (benzoyl peroxide, dicumyl peroxide, and azobisisobutyronitrile) was found to exhibit a negative effect because these initiators interacted with the polymer radicals to form stable polymer, resulting a lower degradation efficiency (Daraboina and Madras 2009). Vinu and Madras (2011) reported a study on sono-photooxidative degradation of PMMA, PEMA, and PBMA in the presence of toluene as solvent and benzoin as photoinitiator. Results showed that the combination of ultrasound + UV + benzoin yielded the highest initial rate of PMMA degradation as compared to other treatments. However, for longer treatment periods, the time evolution profiles of average molecular weight (M_n) for ultrasound + UV + benzoin were similar to those obtained from ultrasound + UV, attaining the same limiting M_n. As for polydispersity profiles, sonophotooxidative degradation resulted in a limiting polydispersity value of 1.6 ± 0.05 for all poly(alkyl methacrylates)s. This value remained in between the limiting values of 1 and 2, which was observed for degradation of the polymer by only ultrasound or UV irradiation, respectively. They concluded that UV + initiator only accelerated the degradation of polymer in the initial period, and once most of the initiator was depleted, UV only acted to break shorter polymer chains, while ultrasound dominated the scission effect of longer chains (Vinu and Madras 2011).

Koda et al. (2011) investigated the effects of frequency and radical scavenger on ultrasonic degradation of four different water-soluble polymers (methyl cellulose, pullulan, dextran, and poly(ethylene oxide)). It was found that the degradation proceeded faster when 500 kHz of ultrasonic frequency was used as compared to 20 kHz. They also confirmed that the presence of t-BuOH could suppress the degradation of these polymers. The highest degradation rate was obtained for methyl cellulose because of the difference in the persistence length and hydrodynamic radius of the polymer (Koda et al. 2011). Desai et al. (2008) studied the effect of different parameters on the degradation of low-density polyethylene in o-dichlorobenzene as solvent using viscometry as a technique for monitoring the rate of degradation. A major extent of the degradation was observed in the initial period of irradiation time. When reaction volume, polymer concentration, or reaction temperature was increased, polyethylene degradation was found to be reduced (Desai et al. 2008). Guo and Peng (2007) presented their study on ultrasonic degradation of polypropylene, which is a semicrystalline polymer with good mechanical and processing properties, chemical resistance, and low density. When ultrasonic irradiation was introduced, a decrease of complex viscosity, zero shear viscosity, representative relaxation time, viscoelastic moduli as well as cross-over modulus and an increase of cross-over frequency were observed due to reduction of average molecular weight (M_w) and increase in molecular weight distribution index (M_w/M_n) of the polymer. From their experiment, chain scission mainly occurred at the initial 3 min of ultrasonic irradiation and subsequently inclined to termination (Guo and Peng 2007). In a study conducted by Mehrdad (2010), the effects of solvent composition and solution concentration on ultrasonic degradation of poly(vinyl pyrrolidone) in aqueous acetone solution were investigated. They showed that the increasing solution concentration or acetone volume fraction would limit the extent of polymer molecular weight reduction by ultrasonic irradiation. It was also found that the rate of degradation and the limiting values of molecular weight of polymer could correlate in terms of the viscosity of polymer solution and vapor pressure of the solvent used (Mehrdad 2011).

With ultrasonic degradation studies focussing on both linear random coils as well as randomly branched polymers, there are still investigations lacking for limiting cases, for example, hard spheres or near-hard spheres at one end of the architectural spectrum, rigid rods or highly extended polymers at the other end.

Studies on the structural effects of polymer architecture on ultrasonic degradation has shown that long-chain branching and conformation can individually influence both the rate and mechanism of degradation (Ostlund et al. 2008). Ostlund et al. (2008) investigated the ultrasonic degradation of polypeptide poly(γ-benzyl-L-glutamate) (PBLG). They found that after 10 h of sonication, molar mass, and size of the polymer decreased to less than half of their original values and a substantial decrease in molar mass polydispersity was observed. However, there was no change in the conformation of the polymer during the degradation. This was found by the invariance in the fractal dimension during sonication, as measured by both light scattering and viscometry, as well as by the lack of change in the ratio of radii obtained by both detection methods (Ostlund et al. 2008).

Aarthi et al. (2007) studied the effect of single or combined treatment using ultrasound and UV irradiation on the degradation of water soluble polymers poly(ethylene oxide), poly(acrylic acid), and poly(vinyl pyrrolidone) in the presence of combustion solution synthesized TiO_2. A higher rate of degradation was observed for all compounds using combined treatment as compared to independent exposure of UV or ultrasound. This was due to the increase in number of scission products per breakage and not due to the increase in the intrinsic rate. A model for degradation was also proposed based on ternary fragmentation, which fitted well with the experimental data for both number-average molecular weight (M_n) and polydispersity (Aarthi et al. 2007).

Traditionally, molecular weight measurements were often a laborious process and were performed by withdrawing several samples periodically from the sonication environment. Akyüz et al. (2008) introduced an "automatic continuous monitoring of polymerization" (ACOMP), which was an online monitoring technique in the study of ultrasonic depolymerisation of poly(vinyl pyrrolidone). Finer details of the process was observed using this technique as the molecular weight of poly(vinyl pyrrolidone) decreased by almost a factor of 20 when subjected to ultrasonic irradiation. The use of ACOMP enabled discrimination among theoretical models to explain the polymer degradation (Akyüz et al. 2008).

3.5 Ultrasound Treatment of Dye Wastewater

Dye is a vital component, which is commonly used in many industries such as textile, cosmetic, paper, leather, pharmaceutical, and food industries (Thangavadivel et al. 2011). Currently, there are over 100,000 commercially available dyes with an estimated overall world dye production of over 7×10^5 ton/year (Grčić et al. 2010a, Chen et al. 2011). However, dye-containing wastewater is causing serious impact on both living organisms and on the environment. From the amount of dyes used during the process, about 1–15 % is discharged to the environment as industrial effluent (Low et al. 2012). Although the concentration of dye is usually lower than any other chemical present, dye often receives the largest attention because the presence of less than 1 mg L^{-1} for some dyes is highly visible and enough to present an esthetic problem (Low et al. 2012). Since the dye is designed to have high resistance to fading caused by chemical, biological, and light-induced, synthetic dyes are generally resistant to oxidative biodegradation (Kritikos et al. 2007). The intense color from dye is able to affect the photosynthetic processes of aquatic plants and reduce the oxygen levels in water, thus affecting the aquatic life (Grčić et al. 2010a). Concerns also arise as many of the dyes are made from known carcinogens such as benzidine and other aromatic compounds (Thangavadivel et al. 2011). For example, non-toxic azo dyes, when present under anaerobic conditions, are cleaved by microorganisms to form potentially carcinogenic aromatic amines (Wang et al. 2011b). Until now, conventional methods such as coagulation, microbial degradation, absorption on activated carbon, incineration, biosorption, filtration, and sedimentation have been used to treat dye wastewater (Chan et al. 2011). There have been considerable interests in the application of ultrasound to remove dye from wastewater (Vajnhandl and Marechal 2007; Guzman-Duque et al. 2011) and some recent studies are summarized in Table 3.5.

3.5.1 Reactive Dyes

Reactive dyes are typically azo-based chromophores, characterized by their azo groups combined with different reactive groups (Hsieh et al. 2009). Classes of reactive dyes include triphendioxazine, phthalocyanine, formazan, and anthraquinone dyes (Hunger 2003). These dyes form covalent bonds with textile fibers, usually cotton, to be colored during its application. Disposal of reactive dyes requires attention owing to its poor biodegradation. Some of the reactive dyes are also toxic and carcinogenic to human beings (Hsieh et al. 2009).

Vajnhandl and Marechal (2007) investigated the extent of decolorization and mineralization of reactive black 5 (RB5) using ultrasonic irradiation without addition of any oxidant. They concluded that ultrasound alone was capable to completely decolorize RB5 and the extent of conversion strongly depended on the operating conditions. Although only 50 % mineralization of dye was achieved after 6 h, no toxic degradation by-products were detected in their study (Vajnhandl and Marechal 2007). Recently, Zhou et al. (2011) studied the role of ligands in the combined ultasonic/UV/Fe^{3+} system to treat RB5 wastewater. In this combined treatment, a synergy factor of 2.5 based on the pseudo-first-order degradation rate constant (k_{obs}) was found, together with the enhancements in organic detoxification and mineralization. Despite the synergistic effect, the relatively slow H_2O_2 production and Fe^{2+} regeneration would limit the OH· formation. The presence of different organic ligands [oxalate, tartrate, succinate, citrate, nitrilotriacetic amine (NTA), and ethyl-enediaminetetra acetic acid (EDTA)] was also found to affect the ultrasonic/UV/Fe^{3+} system. RB5 degradation constant k_{obs}(RB5) followed the sequence of oxalate > tartrate > succinate > citrate > no ligand > N-TA > EDTA. The ligands could be degraded simultaneously with k_{obs}(ligand),

Table 3.5 Degradation performance of dye wastewaters under ultrasonic-enhanced degradation in single/combined treatment

Dye(s)	Treatment(s)	Concentration	Experimental conditions	Results	Reference
Acid Black 1	Ultrasound + Fe^{2+}	0.081 mM (50 mg/L)	40 kHz, 50 W/L, pH 3, 50 mL, 20 °C, 30 min 8 mM H_2O_2, 0.025 mM Fe^{2+}	Decolorization: 98.83 %	Sun et al. (2007)
Acid Blue 25	Ultrasound	50 mg/L	1,700 kHz, 14 W, 399 mg/L CCl_4, pH 5.7, 100 mL, 20 °C, ~20 min	Decolorization initial rate: 15.609 mg L^{-1} min^{-1} Decolorization: ~100 %	Ghodbane and Hamdaoui (2009)
Acid Blue 9 (Brilliant Blue)	Ultrasound + Fe	2×10^{-5} M (~17.1 mg/L)	40 kHz, pH 3, 25 mL, 30 min 2.4 g/L Fe	Decolorization: 93.9 % Mineralization: 6.56 %	Pavanelli et al. (2011)
Acid Orange 10 (Orange-G)	Ultrasound + UV + TiO_2	0.09 mM (40.7 mg/L)	213 kHz, 20 W, pH 5.8, 250 mL, 20 °C, 4 h 450 W Xe-arc lamp, 1 g/L TiO_2 (Degussa P25)	Decolorization initial rate: 22.3×10^{-7} mg L^{-1} min^{-1} TOC degradation: 82 %	Madhavan et al. (2010a)
Acid Orange 3	Ultrasound + fly ash/H_2O_2	100 mg/L	25 kHz, 250 W, pH 3, 50 mL, Room temp. 5.4 mM H_2O_2, 2.5 g/L fly ash	Decolorization: 96 %	Li et al. (2010b)
Acid Orange 5	Ultrasound	100 μM (~37.5 mg/L)	850 kHz, 90 W, 4.1 W/cm^2, 30 °C, 5 h	Decolorization: 100 %	Tauber et al. (2008)
Acid Orange 52	Ultrasound	25 mg/L	200 kHz ultrasonic transducer, 600 W, 50 mL/min air as aeration gas, 1 L, 25 °C	Decolorization: 99 % (240 min) TOC degradation: 30 % (480 min)	Maezawa et al. (2007)
Acid Orange 52	Ultrasound + UV + TiO_2	25 mg/L	200 kHz ultrasonic transducer, 600 W, 50 mL/min air as aeration gas, 1 L, 25 °C Fixed TiO_2, 20 W low pressure mercury lamp (253.6 nm)	Decolorization: 100 % (240 min) TOC degradation: 65 % (480 min)	Maezawa et al. (2007)

(continued)

Table 3.5 (continued)

Dye(s)	Treatment(s)	Concentration	Experimental conditions	Results	Reference
Acid Orange 52 (Methyl Orange)	Ultrasound + UV + TiO$_2$	18 mg/L	47 kHz ultrasonic bath, 81 W, N$_2$ bubbling, pH 6.4, 1.2 L, 25 °C, 60 min 236 mg/L TiO$_2$ (Degussa P25), 4 W black tube lamp (ST fluorescent tube, 300–400 nm), 4.57×10^{-7} Einstein/s	Decolorization: ~95 %	Bejarano-Pérez and Suárez-Herrera (2008)
Acid Orange 52 (Methyl Orange)	Ultrasound + Fe^{2+} + H$_2$O$_2$	100 mg/L	40 kHz ultrasonic bath, 400 W, 400 mL, pH 3, 40 min 12 mM H$_2$O$_2$, 0.5 mM Fe^{2+}	Decolorization: 87.7 %	Chen et al. (2008)
Acid Orange 52 (Methyl Orange)	Ultrasound + UV + TiO$_2$	25 mg/L	20 kHz ultrasonic horn, 240 W, 5 L, pH 4, 35 °C 0.5 mg/L TiO$_2$, 30 W low pressure mercury lamp (254 nm), air bubbling rate: 500 mL/min	Decolorization: ~60 %	Cui et al. (2011)
Acid Orange 52 (Methyl Orange)	Ultrasound + Er^{3+}: YAlO$_3$/TiO$_2$-ZnO	10 mg/L	40 kHz, 50 W, pH 5.5, 25 °C, 120 min 1 g/L Er^{3+}:YAlO$_3$/TiO$_2$-ZnO, 1:1 Ti/Zn molar ratio, 5 wt % Er^{3+}:YAlO$_3$	Decolorization: ~46 %	Gao et al. (2011a)
Acid Orange 52 (Methyl Orange)	Ultrasound + Er^{3+}: YAlO$_3$/TiO$_2$-Fe$_2$O$_3$	10 mg/L	40 kHz, 50 W, pH 5.5, 25 °C, 150 min 1 g/L Er^{3+}:YAlO$_3$/TiO$_2$-Fe$_2$O$_3$, 1:1 Ti/Fe molar ratio	Decolorization: ~50 %	Gao et al. (2011b)
Acid Orange 52 (Methyl Orange)	Ultrasound + TiO$_2$	10 mg/L	40 kHz, 50 W, 22 °C, 180 min 1 g/L TiO$_2$ (Anatase)	Decolorization: ~75 %	Gao et al. (2011c)

(continued)

Table 3.5 (continued)

Dye(s)	Treatment(s)	Concentration	Experimental conditions	Results	Reference
Acid Orange 52 (Methyl Orange)	Ultrasound + electrochemical oxidation + Na$_2$SO$_4$	200 mg/L	59 kHz ultrasonic bath, 90 W, pH 6, 25 °C, 60 min 8 V, three-dimensional electrode reactor, 0.2 mol/L Na$_2$SO$_4$	Decolorization: 99.1 % COD degradation: 83.9 %	He et al. (2011)
Acid Orange 52 (Methyl Orange)	Ultrasound + UV + Ag/TiO$_2$	32 mg/L	40 kHz ultrasonic generator with a transducer, 180 W, pH 6.3, 20 °C, 120 min 36 mg/L Ag/TiO$_2$, 800 W xenon lamp (599 nm)	Decolorization rate constant: 3.58×10^{-2} min^{-1} Decolorization: ~100 %	Wang et al. (2008a)
Acid Orange 52 (Methyl Orange)	Ultrasound + UV + CNTs/TiO$_2$	25 mg/L	20 kHz ultrasonic horn, 50 W, 60 mL 50 mg/L CNTs/TiO$_2$, 30 W black light blue lamp	Decolorization rate constant: 1.118×10^{-2} min^{-1} Decolorization: 66.2 %	Wang et al. (2009b)
Acid Orange 52 (Methyl Orange)	Ultrasound + TiO$_2$	10 mg/L	40 kHz, 50 W, 22 °C, 120 min 1 g/L TiO$_2$	Decolorization rate constant: 4.7×10^{-3} min^{-1} Decolorization: 43.64 %	Wang et al. (2011a)
Acid Orange 52 (Methyl Orange)	Ultrasound + UV + TiO$_2$ + electro-assisted	16.37 mg/L	40 kHz ultrasonic bath, 150 W, 25 °C, 60 min 11 W UV lamp (253.7 nm), TiO$_2$ nanotube electrode and platinum electrode, 0.6 V	Decolorization rate constant: 7.32×10^{-2} min^{-1} Decolorization: ~93 %	Zhang et al. (2008b)
Acid Orange 52 (Methyl Orange)	Ultrasound + Zirconia nanotubes (absorption)	20 mg/L	28 kHz ultrasonic generator, 0.5 W/cm^2, pH 2, 25 °C, 8 h 10 mg/L zirconia nanotubes	Decolorization: 97.6 %	Zhao et al. (2011)
Acid Orange 7	Ultrasound + Er^{3+}: YAlO$_3$/TiO$_2$-ZnO	10 mg/L	40 kHz, 50 W, pH 5.5, 25 °C, 120 min 1 g/L Er^{3+}:YAlO$_3$:YAlO$_3$/TiO$_2$-ZnO, 1:1 Ti/Zn molar ratio, 5 wt % Er^{3+}:YAlO$_3$	Decolorization: ~63 %	Gao et al. (2011a)

(continued)

Table 3.5 (continued)

Dye(s)	Treatment(s)	Concentration	Experimental conditions	Results	Reference
Acid Orange 7	Ultrasound + Fe^0/granular activated carbon (GAC)	1,000 mg/L	40 kHz ultrasonic bath, 100 W, 100 mL, pH 4, 120 g/L Fe^0, 23 g/L GAC	Decolorization rate constant: 3.91×10^{-2} min^{-1}; Decolorization: 80 %; TOC degradation rate constant: 2.02×10^{-2} min^{-1}; TOC degradation: 57 %	Liu et al. (2007)
Acid Orange 7	Ultrasound + O_3	0.91 mM (32 mg/L)	20 kHz ultrasonic generator, 250 W, 201 W/L, 833 mL/min O_2 flow, pH 4.5, 61 °C 0.81 mM O_3	TOC degradation: ~100 %	Zhang et al. (2008a)
Acid Orange 7	Ultrasound + Fe^0 + H_2O_2	200 mg/L	20 kHz ultrasonic generator, 201 W/L, pH 3, 18 °C, 0.5 g/L Fe^0, 15 mM H_2O_2	Decolorization: ~90 % (2 min); COD degradation: 56 % (60 min)	Zhang et al. (2009b)
Acid Orange 7	Ultrasound + Goethite + H_2O_2	79.5 mg/L	20 kHz ultrasonic generator, 80 W/L, 20 °C, 7.77 mM H_2O_2, 0.3 g/L goethite	Decolorization: ~100 % (30 min); TOC degradation: 42 % (90 min)	Zhang et al. (2009a)
Acid Orange 7	Ultrasound + Fe_2O_3-Al_2O_3-meso + H_2O_2	100 mg/L	20 kHz, 80 W, 200 mL, pH 3, 20 °C, 60 min, 0.3 g/L Fe_2O_3-Al_2O_3-meso, 4 mM H_2O_2	Decolorization: ~50 %	Zhong et al. (2011a)
Acid Orange 7	Ultrasound + UV + Fe_2O_3/SBA-15 + H_2O_2	100 mg/L	20 kHz, 80 W, 200 mL, pH 2, 20 °C, 60 min, 4 W UV light tube (254 nm), 0.3 g/L Fe_2O_3/SBA-15, 8 mM H_2O_2	Decolorization: ~80 %	Zhong et al. (2011b)

(continued)

Table 3.5 (continued)

Dye(s)	Treatment(s)	Concentration	Experimental conditions	Results	Reference
Acid Orange 7 (Orange II)	Ultrasound + Au/ TiO$_2$	5×10^{-5} M (16.4 mg/ L)	40 kHz ultrasonic transducer, 50 W, 0.045 W/cm^3, Ar atmosphere, pH 3, 100 mL, 22–28 °C 1 g/L Au/TiO$_2$	Decolorization rate constant: 1.78×10^{-2} s^{-1} Decolorization: 100 % (180 min) TOC degradation: 80 % ($\sim$9 h)	Wang et al. (2008g)
Acid Red 14	Ultrasound + Fe^0	50 mg/L	59 kHz, 100 mL, pH 6.3, 25 °C, 20 min 30 g/L Fe^0	Decolorization: 96.7 % TOC degradation: 43.2 %	Lin et al. (2008)
Acid Red 14 (Acid Red B)	Ultrasound + Er^{3+}: YAlO$_3$/TiO$_2$-ZnO	10 mg/L	40 kHz, 50 W, pH 5.5, 25 °C, 120 min 1 g/L Er^{3+}:YAlO$_3$/TiO$_2$-ZnO, 1:1 Ti/Zn molar ratio, 5 wt % Er^{3+}:YAlO$_3$	Decolorization rate constant: 8.7×10^{-3} min^{-1} Decolorization: 76.54 %	Gao et al. (2011a)
Acid Red 14 (Acid Red B)	Ultrasound + Er^{3+}: YAlO$_3$/TiO$_2$-Fe$_2$O$_3$	10 mg/L	40 kHz, 50 W, pH 5.5, 25 °C, 150 min 1 g/L Er^{3+}:YAlO$_3$/TiO$_2$-Fe$_2$O$_3$, 1:1 Ti/Fe molar ratio	Decolorization rate constant: 7.2×10^{-3} min^{-1} Decolorization: 67.22 %	Gao et al. (2011b)
Acid Red 14 (Acid Red B)	Ultrasound + TiO$_2$ (Anatase, rutile)	10 mg/L	40 kHz, 50 W, 50 mL, 50 cm^2, pH 3, 30 °C, 60 min 1 g/L TiO$_2$	Decolorization: 78 % (Anatase), $\sim$40 % (Rutile)	Wang et al. (2007a)
Acid Red 14 (Acid Red B)	Ultrasound + ZnO	10 mg/L	40 kHz, 50 W, pH 7, 25 °C, 60 min 1 g/L ZnO (nano-sized)	Decolorization rate constant: 1.6×10^{-2} min^{-1} Decolorization: 71.2 %	Wang et al. (2008b)

(continued)

Table 3.5 (continued)

Dye(s)	Treatment(s)	Concentration	Experimental conditions	Results	Reference
Acid Red 14 (Acid Red B)	Ultrasound + ZnO + Oxidants (KClO$_4$, KClO$_3$, Ca(ClO)$_2$)	10 mg/L	40 kHz, 50 W, 50 mL, pH 7, 25 °C, 60 min 1 g/L ZnO, 10 mM oxidant	Decolorization rate constant: 1.86×10^{-2} min^{-1} (Ca(ClO)$_2$), 1.998×10^{-2} min^{-1} (KClO$_3$), 2.09×10^{-2} min^{-1} (KClO$_4$) Decolorization: 68.5 % (Ca(ClO)$_2$), 72.4 % (KClO$_3$), 74.5 % (KClO$_4$)	Wang et al. (2009a)
Acid Red 14 (Acid Red B)	Ultrasound + TiO$_2$	10 mg/L	40 kHz, 50 W, 22 °C, 120 min 1 g/L TiO$_2$	Decolorization rate constant: 1×10^{-2} min^{-1} Decolorization: 70.37 %	Wang et al. (2011b)
Acid Red 97	Ultrasound	50 mg/L	40 kHz, 250 W, 50 mL, pH 3, 40 °C, 140 min	Decolorization rate constant: 5.4×10^{-4} min^{-1}	Li and Song (2010)
Acid Red 97	Ultrasound + Fe^{2+} + H$_2$O$_2$	50 mg/L	40 kHz, 250 W, 50 mL, pH 3, 40 °C, 140 min 1.57 mM H$_2$O$_2$, 0.054 mM Fe^{2+}	Decolorization rate constant: 4.9×10^{-2} min^{-1} Decolorization: 99.9 %	Li and Song (2010)
Acid Violet 19 (Acid fuchsine)	Ultrasound + TiO$_2$	20 mg/L	40 kHz, 50 W, pH 3, 40 °C, 180 min 500 mg/L TiO$_2$ (Rutile)	Decolorization: ~80 %	Wang et al. (2007c)
Acid Violet 19 (Acid fuchsine)	Ultrasound + Fe-doped TiO$_2$	10 mg/L	40 kHz, 50 W, pH 7, 50 mL, 20 °C, 60 min 1 g/L Fe-doped TiO$_2$	Decolorization: ~58.9 %	Wang et al. (2008c)

(continued)

Table 3.5 (continued)

Dye(s)	Treatment(s)	Concentration	Experimental conditions	Results	Reference
Acid Violet 7 (Azo fuchsine)	Ultrasound + Er^{3+}: $YAlO_3/TiO_2$-ZnO	10 mg/L	40 kHz, 50 W, pH 5.5, 25 °C, 120 min 1 g/L Er^{3+}:$YAlO_3/TiO_2$-ZnO, 1:1 Ti/Zn molar ratio, 5 wt % Er^{3+}:$YAlO_3$	Decolorization: ~63 %	Gao et al. (2011a)
Acid Violet 7 (Azo fuchsine)	Ultrasound + Er^{3+}: $YAlO_3/TiO_2$-Fe_2O_3	10 mg/L	40 kHz, 50 W, pH 5.5, 25 °C, 150 min 1 g/L Er^{3+}:$YAlO_3/TiO_2$-Fe_2O_3, 1:1 Ti/Fe molar ratio	Decolorization: ~71 %	Gao et al. (2011b)
Acid Violet 7 (Azo fuchsine)	Ultrasound + TiO_2	10 mg/L	40 kHz, 50 W, 22 °C, 180 min 1 g/L TiO_2 (Anatase)	Decolorization: ~83 %	Gao et al. (2011c)
Acid Violet 7 (Azo fuchsine)	Ultrasound + TiO_2	10 mg/L	40 kHz, 50 W, 22 °C, 120 min 1 g/L TiO_2	Decolorization rate constant: 5.3×10^{-3} min^{-1} Decolorization: 47.58 %	Wang et al. (2011a)
Acid Yellow 17	Ultrasound + UV + TiO_2	12.5 mg/L	10 W, pH 4 50 mL, 250 min 10 W UV lamp, 0.75 g/L TiO_2	Decolorization: ~85 %	Chang et al. (2010)
Basic Blue 41	Ultrasound + TiO_2 + H_2O_2	15 mg/L	35 kHz ultrasonic bath, 160 W, pH 8, 100 mL, 25 °C, 180 min 250 mg/L H_2O_2, 1 g/L TiO_2 nanoparticles	Decolorization: ~85 %	Abbasi and Asl (2008)
Basic Blue 9	Ultrasound + UV + TiO_2	0.06 mM (19.2 mg/L)	20 kHz sonophotoreactor, 76 W, pH 3, 300 mL, 25 °C, 80 min 100 mg/L TiO_2, 15 W UV lamp (352 nm)	Decolorization rate constant: 8.94×10^{-2} min^{-1} Decolorization: ~100 % COD degradation: >80 %	González and Martínez (2008)

(continued)

Table 3.5 (continued)

Dye(s)	Treatment(s)	Concentration	Experimental conditions	Results	Reference
Basic Blue 9	Ultrasound + UV + TiO$_2$ nanotube array film	6 mg/L	27 kHz, 24 W, 100 mL, 3 h 16 W (UV lamp, 365 nm)	Decolorization rate constant: 3.56×10^{-3} min^{-1} Decolorization: ~48 %	Yuan et al. (2009)
Basic Blue 9 (Methylene Blue)	Ultrasound + TiO$_2$	10 mg/L	40 kHz, 50 W, 22 °C, 180 min 1 g/L TiO$_2$ (Anatase)	Decolorization: ~80 %	Gao et al. (2011c)
Basic Blue 9 (Methylene Blue)	Ultrasound + TiO$_2$	2.5×10^{-5} M (~8 mg/L)	176 kHz, 160 W, 20 °C, 90 min 0.2 g/L TiO$_2$ (Anatase), UV-LED	Decolorization: ~93 %	Hayashi et al. (2010)
Basic Blue 9 (Methylene Blue)	Ultrasound + Fe-fullerene/ TiO$_2$	3.2 mg/L	20 kHz ultrasonic processor, 750 W, 50 mL, 35 °C, 150 min 1 g/L Fe-fullerene/TiO$_2$	Decolorization: ~90 %	Meng and Oh (2011)
Basic Blue 9 (Methylene Blue)	Ultrasound + TiO$_2$ + H$_2$O$_2$	0.29 mM (92.8 mg/L)	39 kHz ultrasonic bath, 200 W, pH 7.3, 20 °C, 60 min 50 mM H$_2$O$_2$, 2 kg/L TiO$_2$	Decolorization: ~90 %	Shimizu et al. (2007)
Basic Blue 9 (Methylene Blue)	Ultrasound + TiO$_2$ + H$_2$O$_2$	10 mg/L	40 kHz, 50 W, pH 7, 50 mL, 25 °C, 180 min 1 g/L TiO$_2$	Decolorization: 92.28 %	Wang et al. (2008d)
Basic Blue 9 (Methylene Blue)	Ultrasound + TiO$_2$	10 mg/L	40 kHz, 50 W, 22 °C, 120 min 1 g/L TiO$_2$	Decolorization rate constant: 7.8×10^{-3} min^{-1} Decolorization: 57.43 %	Wang et al. (2011a)
Basic Blue 9 (Methylene Blue)	Ultrasound + TiO$_2$-CNT	1×10^{-5} M (3.2 mg/L)	28 kHz ultrasonic generator, 50 mL, 23 °C,120 min	Decolorization rate constant: 5.32×10^{-3} min^{-1} Decolorization: ~100 %	Zhang et al. (2011b)

(continued)

Table 3.5 (continued)

Dye(s)	Treatment(s)	Concentration	Experimental conditions	Results	Reference
Basic Green (Malachite Green)	Ultrasound	5 mg/L	35 kHz ultrasonic bath, 49 W, 0.16 W/mL, 1 L, 21 °C, 200 min	Decolorization: 100 %	Behnajady et al. (2008b)
Basic Green (Malachite Green)	Ultrasound + UV + H_2O_2	5 mg/L	35 kHz ultrasonic bath, 49 W, 0.049 W/mL, 1 L, 21 °C, 15 min 400 mg/L H_2O_2, 15 W low pressure mercury lamp (UVC, 254 nm), 6 W/m^2	Decolorization rate constant: 3.556×10^{-1} min^{-1} Decolorization: 100 %	Behnajady et al. (2008c)
Basic Green (Malachite Green)	Ultrasound + CCl_4	2.75×10^{-5} M (10 mg/L)	20 kHz ultrasonic probe, 11 W, 0.5 mL/L CCl_4, 1 L, 25 °C,60 min	Decolorization rate constant: sim;4.8×10^{-2} min^{-1}	Bejarano-Pérez and Suárez-Herrera (2008)
Basic Green (Malachite Green)	Ultrasound + UV + TiO_2	2.75×10^{-5} M (10 mg/L)	20 kHz ultrasonic probe, 11 W, 0.5 mL/L CCl_4, 1 L, 25 °C,60 min 0.5 g/L TiO_2 (99 % anatase), black bulb (ST fluorescent tube, 300–400 nm), 4.57×10^{-7} Einstein/s	Decolorization rate constant: $\sim 4.8 \times 10^{-2}$ min^{-1}	Bejarano-Pérez and Suárez-Herrera (2008)
Basic Green (Malachite Green)	Ultrasound + Er^{3+}: $YAlO_3$/TiO_2-Fe_2O_3	10 mg/L	40 kHz, 50 W, pH 5.5, 25 °C, 150 min 1 g/L Er^{3+}:$YAlO_3$/TiO_2-Fe_2O_3, 1:1 Ti/Fe molar ratio	Decolorization: ~ 63 %	Gao et al. (2011b)
Basic Green (Malachite Green)	Ultrasound	10 mg/L	300 kHz piezoelectric disk, 60 W, 10 g/L KBr, pH 5.3, 300 mL, 25 °C, 30 min	Decolorization: 100 %	Moumeni and Hamdaoui (2011)

(continued)

Table 3.5 (continued)

Dye(s)	Treatment(s)	Concentration	Experimental conditions	Results	Reference
Basic Red 29	Ultrasound + H_2O_2 + Co(II)Act	20 mg/L	40 kHz ultrasonic bath, 25 W, 0.25 W/mL, pH 6.7, 40 °C, 30 min 1,000 mg/L H_2O_2, 1,000 mg/L Co(II)Act	Decolorization rate constant: 0.1211 min^{-1} Decolorization: 100 %	Yavuz et al. (2009)
Basic Red 46 (Cationic Red-GRL)	Ultrasound + Fe^{2+} (Electro-Fenton) + Na_2SO_4	150 mg/L	20 kHz, 160 W, pH 3, 100 mL, 27 °C, 180 min 5 mM Fe^{2+}, 8.89 mA/cm^2, 0.05 M Na_2SO_4	TOC degradation: 68.45 %	Li et al. (2010a)
Basic Violet 1 (Methyl Violet)	Ultrasound + ZnO + Oxidants ($KClO_4$, $KClO_3$, $Ca(ClO_2)_2$)	10 mg/L	40 kHz, 50 W, 50 mL, pH 7, 25 °C, 60 min 1 g/L ZnO, 10 mM oxidant	Decolorization: ~39 % ($Ca(ClO)_2$), ~44 % ($KClO_3$), ~52 % ($KClO_4$)	Wang et al. (2009a)
Basic Violet 10 (Rhodamine B)	Ultrasound	5 mg/L	35 kHz ultrasonic system, 49 W, 329 cm^2, 0.163 W/mL, 25 °C, 60 min	Decolorization: ~87 %	Behnajady et al. (2008a)
Basic Violet 10 (Rhodamine B)	Ultrasound + Er^{3+} : $YAlO_3/TiO_2$-Fe_2O_3	10 mg/L	40 kHz, 50 W, pH 5.5, 25 °C, 150 min 1 g/L Er^{3+}:$YAlO_3/TiO_2$-Fe_2O_3, 1:1 Ti/Fe molar ratio	Decolorization: ~66 %	Gao et al. (2011b)
Basic Violet 10 (Rhodamine B)	Ultrasound + FeO + H_2O_2	5.69×10^{-5} M (27.26 mg/L)	24 kHz ultrasonic generator, 400 W, 25 °C, ~50 min 5.815 M H_2O_2, 2 g/L FeO	Decolorization rate constant: 2.97×10^{-2} min^{-1} Decolorization: ~77 % (140 min)	Mehrdad and Hashemzadeh (2010)
Basic Violet 10 (Rhodamine B)	Ultrasound	5 mg/L	300 kH, 60 W, pH 5.3, 300 mL, 25 °C	Decolorization: 100 % (140 min) COD degradation: 58 % (240 min)	Merouani et al. (2010a)

(continued)

Table 3.5 (continued)

Dye(s)	Treatment(s)	Concentration	Experimental conditions	Results	Reference
Basic Violet 10 (Rhodamine B)	Ultrasound + bicarbonate	0.5 mg/L	300 kH, 60 W, 3 g/L bicarbonate, pH 8.3, 300 mL, 25 °C	Decolorization: 100 % (25 min); COD degradation: 100 % (60 min)	Merouani et al. (2010b)
Basic Violet 10 (Rhodamine B)	Ultrasound + carbonate	0.5 mg/L	300 kH, 60 W, 3 g/L carbonate, pH 11.1, 300 mL, 25 °C	Decolorization: 100 % (15 min); COD degradation: 100 % (40 min)	Merouani et al. (2010b)
Basic Violet 10 (Rhodamine B)	Ultrasound + UV + TiO_2	10 mg/L	25 kHz ultrasonic bath, 195 W, pH 2.5, 30 °C, 180 min 2 g/L TiO_2	Decolorization: 92.9 % TOC degradation: 50 %	Mishra and Gogate (2011b)
Basic Violet 10 (Rhodamine B)	Ultrasound + TiO_2 + H_2O_2	50 mg/L	35 kHz ultrasonic bath, 50 W, air 1L/min, pH 6, 200 mL, 30 °C, 120 min 2 g/L TiO_2 nanotubes, 8 mM H_2O_2	Decolorization rate constant: 1.5×10^{-2} min^{-1} Decolorization: 85 %	Pang et al. (2011b)
Basic Violet 10 (Rhodamine B)	Ultrasound + TiO_2	44.8 mg/L	35 kHz ultrasonic bath, 68.9 W, 200 mL, 30 °C, 3 h 2.14 g/L TiO_2 nanotubes	Decolorization rate constant: 1.6×10^{-2} min^{-1} Decolorization: 94.6 %	Pang et al. (2011a)
Basic Violet 10 (Rhodamine B)	Ultrasound + ZnO	10 mg/L	40 kHz, 50 W, pH 7, 25 °C, 60 min 1 g/L ZnO (nano-sized)	Decolorization rate constant: 6.6×10^{-3} min^{-1} Decolorization: 39.1 %	Wang et al. (2008b)
Basic Violet 10 (Rhodamine B)	Ultrasound	20 mg/L	40 kHz, 300 W, 2.93×10^{-7} mg/J, pH 5.4, 100 mL, 3 h	Decolorization: 47.6 %	Wang et al. (2008e)

(continued)

Table 3.5 (continued)

Dye(s)	Treatment(s)	Concentration	Experimental conditions	Results	Reference
Basic Violet 10 (Rhodamine B)	Ultrasound + ZnO + Oxidants (KClO$_4$, KClO$_3$, Ca(ClO)$_2$)	10 mg/L	40 kHz, 50 W, 50 mL, pH 7, 25 °C, 60 min, 1 g/L ZnO, 10 mM oxidant	Decolorization: ~27 % (Ca(ClO)$_2$), ~21 % (KClO$_3$), ~25 % (KClO$_4$)	Wang et al. (2009a)
Basic Violet 10 (Rhodamine B)	Ultrasound + Fe$_3$O$_4$ + H$_2$O$_2$	0.02 mM (9.58 mg/L)	20 kHz ultrasonic system, 6 W, pH 5, 25 °C, 60 min, 40 mM H$_2$O$_2$, 0.5 g/L Fe$_3$O$_4$	Decolorization rate constant: 3.5×10^{-2} min^{-1} Decolorization: 90 %	Wang et al. (2010b)
Basic Violet 10 (Rhodamine B)	Ultrasound + TiO$_2$	10 mg/L	40 kHz, 50 W, 22 °C, 120 min, 1 g/L TiO$_2$	Decolorization rate constant: 2.9×10^{-3} min^{-1} Decolorization: 31.29 %	Wang et al. (2011a)
Basic Violet 3 (Crystal Violet)	Ultrasound	20 mg/L	800 kHz piezoelectric disc, 80 W, pH, 1 mmol/L FeSO$_4$, air as saturating gas, 250 mL, 20 °C, 360 min	Decolorization: 100 % TOC degradation: ~12 %	Guzman-Duque et al. (2011)
Direct Black 168	Ultrasound + fly ash + H$_2$O$_2$	100 mg/L	40 kHz ultrasonic cleaner, 250 W, pH 3, 90 min, 2 g/L fly ash, 2.94 mM H$_2$O$_2$	Decolorization: 99 %	Song and Li (2009)
Direct Blue 15 (Direct Sky Blue 5B)	Ultrasound + Fe^0	20 mg/L	20 kHz, 400 W, 50 mL, pH 3, 25 °C, 20 min, 2 g/L iron powder	Decolorization: 99 %	Chen et al. (2011)
Direct Blue 71	Ultrasound	100 μM (~96.6 mg/L)	850 kHz, 90 W, 4.1 W/cm^2, 30 °C, 23 h	Decolorization: 100 %	Tauber et al. (2008)

(continued)

Table 3.5 (continued)

Dye(s)	Treatment(s)	Concentration	Experimental conditions	Results	Reference
Direct Red 23	Ultrasound + O_3	100 mg/L	20 kHz ultrasonic processor, 400 W, 176 W/L, pH 8, 500 mL, 25 °C, 1 min 3.2 g/h O_3	Decolorization: 98 %	Song et al. (2007)
Direct Red 28 (Congo Red)	Ultrasound + UV + TiO_2 (99 % anatase)	4×10^{-5} M (27.87 mg/L)	47 kHz ultrasonic bath, 81 W, pH 6.4, 1.2 L, 25 °C, ~262 min 460 mg/L TiO_2, 4 W black tube lamp (ST fluorescent tube, 300–400 nm), 4.57×10^{-7} Einstein/s	Decolorization rate constant: $\sim 1.43 \times 10^{-4}$ s^{-1}	Bejarano-Pérez and Suárez-Herrera (2008)
Direct Red 28 (Congo Red)	Ultrasound + Er^{3+}: $YAlO_3/TiO_2$-ZnO	10 mg/L	40 kHz, 50 W, pH 5.5, 25 °C, 120 min 1 g/L Er^{3+}:$YAlO_3/TiO_2$-ZnO, 1:1 Ti/Zn molar ratio, 5 wt % Er^{3+}:$YAlO_3$	Decolorization: ~94 %	Gao et al. (2011a)
Direct Red 28 (Congo Red)	Ultrasound + Er^{3+}: $YAlO_3/TiO_2$-Fe_2O_3	10 mg/L	40 kHz, 50 W, pH 5.5, 25 °C, 150 min 1 g/L Er^{3+}:$YAlO_3/TiO_2$-Fe_2O_3, 1:1 Ti/Fe molar ratio	Decolorization: ~44 %	Gao et al. (2011b)
Direct Red 28 (Congo Red)	Ultrasound	58.6 mg/L	30 kHz sonochemical reactor, pH 4, 3 h	Decolorization: 95.9 %	Gopinath et al. (2010)
Direct Red 28 (Congo Red)	Ultrasound + TiO_2	10 mg/L	40 kHz, 50 W, 50 mL, pH 3–5, 50 °C, 180 min 1–1.5 g/L TiO_2 (nanometer rutile)	Decolorization: 100 %	Wang et al. (2007b)

(continued)

Table 3.5 (continued)

Dye(s)	Treatment(s)	Concentration	Experimental conditions	Results	Reference
Direct Red 28 (Congo Red)	Ultrasound	10 mg/L	40 kHz, 50 W, 50 mL, pH 7, 25 °C, 60 min	Decolorization: ~35 %	Wang et al. (2009a)
Direct Red 28 (Congo Red)	Ultrasound + ZnO + Oxidant (KClO$_4$, KClO$_3$, Ca(ClO)$_2$)	10 mg/L	40 kHz, 50 W, 50 mL, pH 7, 25 °C, 60 min 1 g/L ZnO, 10 mM oxidant	Decolorization: ~93 % (Ca(ClO)$_2$), ~98 % (KClO$_3$), ~100 % (KClO$_4$)	Wang et al. (2009a)
Direct Red 28 (Congo Red)	Ultrasound + TiO$_2$	10 mg/L	40 kHz, 50 W, 22 °C, 120 min 1 g/L TiO$_2$	Decolorization rate constant: 2.5×10^{-3} min^{-1} Decolorization: 25.69 %	Wang et al. (2011a)
Direct Yellow 12	Ultrasound + Er^{3+}: YAlO$_3$/TiO$_2$-ZnO	10 mg/L	40 kHz, 50 W, pH 5.5, 25 °C, 120 min 1 g/L Er^{3+}:YAlO$_3$/TiO$_2$-ZnO, 1:1 Ti/Zn molar ratio, 5 wt % Er^{3+}:YAlO$_3$	Decolorization: ~50 %	Gao et al. (2011a)
Direct Yellow 9	Ultrasound	28.75 µM (~20 mg/L)	577 kHz, 48.9 W, 0.46 W/mL, pH 6.9, 30 min	Decolorization: 22.58 %	Eren and Ince (2010)
Disperse Orange 25	Ultrasound	10 mg/L	42 kHz ultrasonic generator, 170 W, 2.5 L, 300 min	Decolorization: ~11 %	Maleki et al. (2010)
Food Red 17 (Red 40)	Ultrasound + Fe	2×10^{-5} M (~10 mg/L)	40 kHz, pH 3, 25 mL, 30 min 2.4 g/L Fe	Decolorization: 99.86 % Mineralization: 7.8 %	Pavanelli et al. (2011)
Food Red 9 (Amaranth)	Ultrasound + Fe	2×10^{-5} M (~12.1 mg/L)	40 kHz, pH 3, 25 mL, 30 min 2.4 g/L Fe	Decolorization: 91.63 % Mineralization: 7.7 %	Pavanelli et al. (2011)
Food Red 9 (Amaranth)	Ultrasound + La^{3+} doped TiO$_2$	10 mg/L	40 kHz, 80 W, pH 7, 100 mL, 90 min 1 g/L La^{3+} doped TiO$_2$	Decolorization: ~100 %	Song et al. (2011b)

(continued)

Table 3.5 (continued)

Dye(s)	Treatment(s)	Concentration	Experimental conditions	Results	Reference
Food Red 9 (Amaranth)	Ultrasound + Tb_7O_{12}/TiO_2	10 mg/L	40 kHz, 80 W, pH 7, 100 mL, 45 min 1 g/L Tb_7O_{12}/TiO_2	Decolorization: ~100 %	Song et al. (2011a)
Food Yellow 3 (Sunset Yellow)	Ultrasound + Fe	2×10^{-5} M ($\sim$9 mg/L)	40 kHz, pH 3, 25 mL, 30 min 2.4 g/L Fe	Decolorization: 99.53 % Mineralization: 28.28 %	Pavanelli et al. (2011)
Mordant Yellow 10	Ultrasound + Fe^{2+} + H_2O_2 + $S_2O_8^{2-}$	50 mg/L	40 kHz (220 V, 50 Hz, 7 A) ultrasonic bath, pH 3, 100 mL, 25 °C, 60 min 2 mM Fe^{2+}, 5 mM H_2O_2, 5 mM $S_2O_8^{2-}$	Decolorization: ~100 % Mineralization: ~65 %	Grčić et al. (2010a)
Mordant Yellow 10 + Reactive Violet 2	Ultrasound	50 mg/L (each dye)	40 kHz (220 V, 50 Hz, 7 A) ultrasonic bath, pH 3, 100 mL, 25 °C, 60 min	TOC degradation: ~10 %	Grčić et al. (2010a)
Mordant Yellow 10 + Reactive Violet 2	Ultrasound + Fe^{2+} + H_2O_2 + $S_2O_8^{2-}$	50 mg/L (each dye)	40 kHz (220 V, 50 Hz, 7 A) ultrasonic bath, pH 3, 100 mL, 25 °C, 60 min 2 mM Fe^{2+}, 5 mM H_2O_2, 5 mM $S_2O_8^{2-}$	Decolorization: ~100 % Mineralization: ~64 %	Grčić et al. (2010a)
Mordant Yellow 10 + Reactive Violet 2 + Reactive Yellow 3	Ultrasound + Fe^{2+} + H_2O_2 + $S_2O_8^{2-}$	50 mg/L (each dye)	40 kHz (220 V, 50 Hz, 7 A) ultrasonic bath, pH 3, 100 mL, 25 °C, 60 min 2 mM Fe^{2+}, 5 mM H_2O_2, 5 mM $S_2O_8^{2-}$	Decolorization: ~100 % Mineralization: ~73 %	Grčić et al. (2010a)
Mordant Yellow 10 + Reactive Yellow 3	Ultrasound + Fe^{2+} + H_2O_2 + $S_2O_8^{2-}$	50 mg/L (each dye)	40 kHz (220 V, 50 Hz, 7 A) ultrasonic bath, pH 3, 100 mL, 25 °C, 60 min 2 mM Fe^{2+}, 5 mM H_2O_2, 5 mM $S_2O_8^{2-}$	Decolorization: ~100 % Mineralization: ~60 %	Grčić et al. (2010a)

(continued)

Table 3.5 (continued)

Dye(s)	Treatment(s)	Concentration	Experimental conditions	Results	Reference
Reactive Black 5	Ultrasound + nanoscale Fe + Fe^{2+} + H_2O_2	500 mg/L	20 kHz sonicator, Fe dose 1 g/L, [Dye]:$[H_2O_2]$:$[Fe^{2+}]$ = 1:3.6:2.4, 250 mL, pH 2, 25 °C, ~60 min	Decolorization: 99.91 % COD removal: 63.36 %	Hsieh et al. (2009)
Reactive Black 5	Ultrasound	20 mg/L	80 kHz horn type generator, 135 W, Ar purging, 350 mL, pH 7, 25 °C, 120 min	Decolorization: 45 %	Kritikos et al. (2007)
Reactive Black 5	Ultrasound + UV + Hombicat UV 100 TiO_2	60 mg/L	80 kHz horn type generator, 135 W, O_2 sparging, 350 mL, pH 7, 25 °C, 40 min 0.25 g/L TiO_2, UVA irradiation using 9 W UVA lamp (350–400 nm, 4.69×10^{-6} Einstein/s)	Decolorization: ~100 %	Kritikos et al. (2007)
Reactive Black 5	Ultrasound	10 mg/L	42 kHz ultrasonic generator, 170 W, 2.5 L, 300 min	Decolorization: ~34 %	Maleki et al. (2010)
Reactive Black 5	Ultrasound	100 µM (~99.2 mg/L)	850 kHz, 90 W, 4.1 W/cm^2, 30 °C, 9 h	Decolorization: 100 %	Tauber et al. (2008)
Reactive Black 5	Ultrasound	20 mg/L	817 kHz ultrasonic transducer with active acoustic vibration surface area of 25 cm^2, 100 W, Ar as saturating gas, pH 7, 25 °C, ~6 h	Decolorization: 98.7 % TOC degradation: 50 %	Vajnhandl and Marechal (2007)

(continued)

Table 3.5 (continued)

Dye(s)	Treatment(s)	Concentration	Experimental conditions	Results	Reference
Reactive Black 5	Ultrasound +UV + Fe^{3+} + oxalate	20 mg/L	20 kHz sonicator, 300 W, 0.5 mM Fe^{3+}, 1.0 mM oxalate, 1 L/min purified air, 350 mL, pH 3, 60 min; UVA irradiation using 9 W UVA lamp (365 nm)	TOC degradation: 82 %; Final EC_{50}: 197 %	Zhou et al. (2011)
Reactive Blue 19	Ultrasound + O_3	500 mg/L	20 kHz ultrasonic processor, 88 W/L, pH 8, 25 °C, 120 min; 3.6 g/h O_3	TOC reduction rate: 8.2×10^{-3} min^{-1}; TOC degradation: ~60 %	He et al. (2008)
Reactive Blue 19	Ultrasound + O_3	500 mg/L	20 kHz ultrasonic processor, 88 W/L, pH 8, 25 °C, 120 min; 3.6 g/h O_3	TOC reduction rate: 8.2×10^{-3} min^{-1}; TOC degradation: ~60 %	He et al. (2008)
Reactive Blue 4	Ultrasound	70 mg/L	35 kHz ultrasonic bath, 50 W, 200 mL, pH 4, ~60 min	Decolorization: 12 %	Jamalluddin and Abdullah (2011)
Reactive Blue 4	Ultrasound + 0.4Fe(III)/TiO_2	70 mg/L	35 kHz ultrasonic bath, 50 W, 1.5 g/L 0.4Fe(III)/TiO_2 with aeration, 200 mL, pH 4, ~60 min	Decolorization: 96 %	Jamalluddin and Abdullah (2011)
Reactive Brilliant Red K-BP	Ultrasound	10 mg/L	20 kHz ultrasonic reactor, 150 W, 100 mL, pH 3, 20 °C, 275 min	Decolorization rate constant: 1.52×10^{-3} min^{-1}; Decolorization: ~30 %	Wang et al. (2008f)
Reactive Brilliant Red K-BP	Ultrasound + Fe^{2+} + H_2O_2	10 mg/L	20 kHz ultrasonic reactor, 150 W, 20 µmol/L H_2O_2, 5 µmol/L Fe^{2+}, 100 mL, pH 3, 20 °C, 275 min	Decolorization rate constant: 6.01×10^{-3} min^{-1}; Decolorization: ~75 %	Wang et al. (2008f)

(continued)

Table 3.5 (continued)

Dye(s)	Treatment(s)	Concentration	Experimental conditions	Results	Reference
Reactive Orange 16	Ultrasound	100 μM (~61.8 mg/L)	850 kHz, 90 W, 4.1 W/cm^2, 30 °C, 23 h	Decolorization: 100 %	Tauber et al. (2008)
Reactive Red 120	Ultrasound	100 mg/L	47 kHz ultrasonic bath, 130 W, pH 4.1, 6 h	Decolorization: ~75 %	Kavitha and Palanisamy (2011)
Reactive Red 120	Ultrasound + UV + TiO$_2$	100 mg/L	47 kHz ultrasonic bath, 130 W, pH 4.1, 6 h 2.5 gm/L TiO$_2$, 50 W halogen lamp (Philips)	Decolorization: ~94 %	Kavitha and Palanisamy (2011)
Reactive Red 141	Ultrasound	28.75 μM (~56.6 mg/L)	577 kHz, 48.9 W, 0.46 W/mL, pH 6.6, 30 min	Decolorization: 45.73 %	Eren and Ince (2010)
Reactive Red 198	Ultrasound + TiO$_2$	50 mg/L	47 kHz ultrasonic bath, 130 W, 300 mg/L TiO$_2$, pH 4.6, 25 °C, ~180 min	Decolorization rate constant: 3.8 × 10^{-3} h^{-1} Decolorization: ~55 %	Kaur and Singh (2007)
Reactive Red 198	Ultrasound + UV + TiO$_2$	50 mg/L	47 kHz ultrasonic bath, 130 W, pH 4.6, 25 °C, ~180 min 300 mg/L TiO$_2$, 50 W halogen lamp (Philips)	Decolorization rate constant: 2.3 × 10^{-2} h^{-1} Decolorization: ~98 %	Kaur and Singh (2007)
Reactive Red 198	Ultrasound + TiO$_2$	50 mg/L	47 kHz ultrasonic bath, 130 W, 300 mg/L TiO$_2$, pH 4.6, 25 °C, ~180 min	Decolorization rate constant: 3.8 × 10^{-3} h^{-1} Decolorization: ~55 %	Kaur and Singh (2007)
Reactive Red 198	Ultrasound + UV + TiO$_2$	50 mg/L	47 kHz ultrasonic bath, 130 W, pH 4.6, 25 °C, ~180 min 300 mg/L TiO$_2$, 50 W halogen lamp (Philips)	Decolorization rate constant: 2.3 × 10^{-2} h^{-1} Decolorization: ~98 %	Kaur and Singh (2007)

(continued)

Table 3.5 (continued)

Dye(s)	Treatment(s)	Concentration	Experimental conditions	Results	Reference
Reactive Red 2	Ultrasound + UV + TiO$_2$	20 mg/L	40 kHz ultrasonic bath, 400 W, 3 L hollow cylindrical glass reactor, pH 7, 30 °C, 120 min 2 g/L TiO$_2$, 15 W UVC lamp (254 nm, 10 mW/cm^2, Philips)	Decolorization: ~80 % TOC degradation: 63 %	Wu (2009)
Reactive Red 24 (Reactive brilliant Red K-2BP)	Ultrasound	20 mg/L	20 kHz ultrasonic reactor, 90 W, 100 mL, pH 5.5, 50 °C, ~180 min	Decolorization: ~7.5 % Cavitational yield: 1.70×10^{-7} mg/J	Wang et al. (2011b)
Reactive Red 24 (Reactive brilliant Red K-2BP)	Ultrasound + H$_2$O$_2$	20 mg/L	20 kHz ultrasonic reactor, 90 W, 300 mg/L H$_2$O$_2$, 100 mL, pH 5.5, 50 °C, 120 min	Decolorization: ~17.5 % Cavitational yield: 5.25×10^{-7} mg/J	Wang et al. (2011b)
Reactive Red 24 (Reactive brilliant Red K-2BP)	Ultrasound + CCl$_4$	20 mg/L	20 kHz ultrasonic generator, 200 W (Calorimetric power: 17.3 W), 0.03 mL CCl$_4$, 100 mL, pH 3, 20 °C, 50 min	Decolorization: ~100 % COD removal: ~60 %	Wang et al. (2011c)
Reactive Violet 2	Ultrasound	50 mg/L	40 kHz (220 V, 50 Hz, 7 A) ultrasonic bath, pH 3, 100 mL, 25 °C, 60 min	TOC degradation: <5 %	Grčić et al. (2010a)
Reactive Violet 2	Ultrasound + Fe^{2+} + H$_2$O$_2$ + S$_2$O$_8^{2-}$	50 mg/L	40 kHz (220 V, 50 Hz, 7 A) ultrasonic bath, pH 3, 100 mL, 25 °C, 60 min 2 mM Fe^{2+}, 5 mM H$_2$O$_2$, 5 mM S$_2$O$_8^{2-}$	Decolorization: ~100 % Mineralization: ~51 %	Grčić et al. (2010a)

(continued)

Table 3.5 (continued)

Dye(s)	Treatment(s)	Concentration	Experimental conditions	Results	Reference
Reactive Violet 2 + Reactive Yellow 3	Ultrasound + Fe^{2+} + H_2O_2 + $S_2O_8^{2-}$	50 mg/L (each dye)	40 kHz (220 V, 50 Hz, 7 A) ultrasonic bath, pH 3, 100 mL, 25 °C, 60 min; 2 mM Fe^{2+}, 5 mM H_2O_2, 5 mM $S_2O_8^{2-}$	Decolorization: ~100 %; Mineralization: ~49 %	Grčić et al. (2010a)
Reactive Yellow	Ultrasound	5 mg/L	130 kHz ultrasonic generator with two piezoelectric transducers, 500 W, 120 min	Decolorization: ~81 %	Dehghani et al. (2008)
Reactive Yellow	Ultrasound	5 mg/L	130 kHz ultrasonic generator with two piezoelectric transducers, 500 W, 120 min	Decolorization: ~81 %	Dehghani et al. (2008)
Reactive Yellow 3	Ultrasound	50 mg/L	40 kHz (220 V, 50 Hz, 7 A) ultrasonic bath, pH 3, 100 mL, 25 °C, 60 min	TOC degradation: <5 %	Grčić et al. (2010a)
Reactive Yellow 3	Ultrasound + Fe^{2+} + H_2O_2 + $S_2O_8^{2-}$	50 mg/L	40 kHz (220 V, 50 Hz, 7 A) ultrasonic bath, pH 3, 100 mL, 25 °C, 60 min; 2 mM Fe^{2+}, 5 mM H_2O_2, 5 mM $S_2O_8^{2-}$	Decolorization: ~100 %; Mineralization: ~48 %	Grčić et al. (2010a)
Reactive Yellow 3	Ultrasound	50 mg/L	40 kHz (220 V, 50 Hz, 7 A) ultrasonic bath, pH 3, 100 mL, 25 °C, 60 min	TOC degradation: <5 %	Grčić et al. (2010a)
Reactive Yellow 3	Ultrasound + Fe^{2+} + H_2O_2 + $S_2O_8^{2-}$	50 mg/L	40 kHz (220 V, 50 Hz, 7 A) ultrasonic bath, pH 3, 100 mL, 25 °C, 60 min; 2 mM Fe^{2+}, 5 mM H_2O_2, 5 mM $S_2O_8^{2-}$	Decolorization: ~100 %; Mineralization: ~48 %	Grčić et al. (2010a)

(continued)

Table 3.5 (continued)

Dye(s)	Treatment(s)	Concentration	Experimental conditions	Results	Reference
Reactive Yellow 84	Ultrasound	500 mg/L	20 kHz, 44 W/L, pH 10, 25 °C, 60 min	TOC degradation: 9 %	He et al. (2007a)
Reactive Yellow 84	Ultrasound + O_3	500 mg/L	20 kHz, 44 W/L, pH 10, 25 °C, 60 min, 4.5 g/h O_3	TOC degradation: 56 %	He et al. (2007a)

following the order of oxalate > citrate > tartrate > succinate > NTA > EDTA. Among the investigated ligands, the presence of oxalate showed the highest increase of mineralization rate of RB5 because the activated oxalate radicals $(C_2O_4 \cdot ^-)$ could be reduced to $CO_2 \cdot ^-/CO_2$ during photo-generation reactions of H_2O_2 (Zhou et al. 2011). Kritikos et al. (2007) investigated the combination of ultrasonic with photocatalysis using Hombicat UV 100 to treat RB5 and showed that the efficiency of sonocatalytic decolorization was more significant at higher dye concentration (60 mg/L as compared to 20 mg/L). This combination was far more effective (~ 100 % for 60 mg/L RB5) as compared with sonolysis alone, which resulted only 45 % decolorization of 20 mg/L RB5 (Kritikos et al. 2007). Grčić et al. (2010a) studied the kinetics of dye mineralization and decolorization of different combinations of dyes (RB5, Mordant Yellow 10 and Reactive Violet 2) using advanced Fenton degradation, $Fe^{2+}/H_2O_2/S_2O_8^{2-}$, with and without ultrasound. TOC content was reduced in higher extent with the presence of ultrasound although ultrasound alone resulted in low mineralization. Besides producing more radicals, an introduction of ultrasound into the process allowed better homogenization of the system which resulted in better mass transfer. Interestingly, ultrasound gave the highest TOC degradation (30 %) in system containing all three different dyes as compared to a single dye system (<10 %). This was explained using the theory of cavitation where for relatively low concentrations, an increase in concentration of dissolved dyes will result in higher intensity of cavitation due to increasing cavitational threshold (Grčić et al. 2010a). Other RB5 treatment studies include optimization of treatment conditions using Taguchi approach on the combination of ultrasound/Fenton/nanoscale iron process, yielding a very high decolorization (99.91 %) (Hsieh et al. 2009).

Wang et al. (2011c) showed that ultrasonic degradation of reactive brilliant red K-2BP (K-2BP) could be enhanced though addition of CCl_2. This enhancement was due to the increase of $OH \cdot$ concentration as CCl_2 acts as a hydrogen scavenger and forms oxidizing agents such as free chlorine and chlorine-containing radicals (Wang et al. 2011c). Wang et al. (2008f) investigated the degradation of reactive brilliant red K-BP (K-BP) and found that it followed pseudo-first-order reaction kinetics and could be accelerated by the addition of Fe^{2+}, Fenton reagent, or NaCl. For example, an addition of 1.5 mol/L of NaCl would result in a 68.9 % increase in degradation rate constant as compared to a system without any addition of NaCl. An addition of NaCl could result in an increase in hydrophilicity, surface tension, and ionic strength of the aqueous phase but a decrease of vapor pressure, thus causing a more violent collapsing of bubbles (Wang et al. 2008f).

For a degradation of Reactive Blue 4 (RB4), Jamalluddin and Abdullah (2011) investigated the combination of $Fe(III)/TiO_2$ catalyst with ultrasonic irradiation and successfully obtained a removal efficiency of 90 %. This combination was further improved (~ 96 %) by introducing an aeration because the splitting of O_2 during the sonication of the solution would generate $O \cdot$ which could subsequently react with the H_2O molecules to form additional $OH \cdot$ (Jamalluddin and Abdullah 2011). Wu (2009) studied the effect of adding $Na_2S_2O_8$ into ultrasound/UV/TiO_2 system to treat Reactive Red 2 (RR2). He found that the presence of persulfate ions was able to trap the photogenerated electrons, preventing their recombination. This phenomenon would simultaneously encourage the generation of sulfate-free radicals, leading to higher decolorization rate as compared to normal ultrasound/UV/ TiO_2 system. He et al. (2007a) conducted a lab-scale experiment to determine the extent of mineralization of Reactive Yellow 84 using sonolytic-ozonation (O_3/ ultrasound) system. The rate constant obtained from the combined system was 18 % higher than the value obtained for the linear combination of ozonation or sonolysis alone, indicating a synergistic increase in overall degradation. The combined system improved the mass transfer of ozone and enhanced the generation of excess radical species. In order to demonstrate the higher production of OH· amount using this combined process, succinic acid was used as test compound as it can only be oxidized by OH· rather than molecular O_3. Results showed that rate constant obtained using ultrasound/O_3 was 119 % higher as compared to its linear combination. They concluded that the formation of additional free radicals was caused by the decomposition of O_3 by sonication (He et al. 2007b).The efficiency of combined ultrasound/O$_{-3}$ was also tested on Reactive Blue 19 (RB19). This system exhibited higher mineralization rate as compared to ozonation alone and could degrade the carbon elements in RB19 into organic acids (such as oxalate acid or acetate acid) with smaller molecular weight (He et al. 2008).

3.5.2 Basic Dyes

Basic dyes are water soluble cationic dyes that are commonly applied to paper, polyacrylonitrile, modified nylons, and modified polyesters (Hunger 2003). For example, Basic Violet 10, also known as Rhodamine B (RhB), is an important xanthene class basic dye which is widely used as a colorant in textile, leather, jute, and food industries and also as a water tracer fluorescent (Merouani et al. 2010a, Pang et al. 2011b). However, its carcinogenicity, reproductive, and developmental toxicities as well as neurotoxicity and chronic toxicities toward humans and animals have been experimentally proven (Jain et al. 2007). Thus, it must be removed before reaching the waterways. Ultrasonic treatment in the presence of iron oxide and H_2O_2 on the degradation of RhB was studied by Mehrdad and Hashemzadeh (2010). The use of ultrasound was found to be aiding the recycling of Fe^{2+}, which in turn accelerated the rate of RhB degradation. A major product of the degradation was 1,2-Benzenedicarboxylic acid diethyl ester (Mehrdad and Hashemzadeh 2010). Sonolysis of RhB was found to follow two different pathways; (1) cleavage of aromatic chromophore ring structure and (2) N-deethylation, in which the former is usually a predominant pathway (Behnajady et al. 2008a).

Merouani et al. (2010a) showed that the degradation of RhB using ultrasonic treatment did not follow first order kinetic. They concluded that the degradation of dye took place at bubble/solution interface. A heterogeneous kinetics model based on a Langmuir-type mechanism was applied to explain the local reaction zone in the interfacial region of cavitation bubbles. Effects of additives such as iron, carbon tetrachloride, H_2O_2, tert-butyl alcohol, salt, sucrose, and glucose on so-nochemical degradation of RhB were also investigated (Merouani et al. 2010a). In their other study, Merouani et al. (2010b) found that the effectiveness of RHB sonolysis was intensified in the presence of bicarbonate or carbonate ions, espe-cially at lower dye concentrations. Similar to $Br_2 \cdot^-$ radicals (Moumeni and Hamdaoui 2011), the generated carbonate radicals would undergo a radical–rad-ical recombination at a lesser extent than OH· and were able to migrate far away from cavitation bubbles for decomposition of RhB molecules in bulk solution. They also found that large concentrations of organic competitor (such as glucose) in the presence of bicarbonate or carbonate ions would decrease the rate of RhB degradation (Merouani et al. 2010b).

Wang et al. (2010b) reported an interesting study on incorporating ultrasound-H_2O_2 system with Fe_3O_4 magnetic nanoparticles (Fe_3O_4-MNP) in the treatment of RhB. No degradations were observed for treatment using H_2O_2 or ultrasound alone. However, the combination of ultrasound-H_2O_2-Fe_3O_4 successfully yielded a 90 % degradation of RhB within 60 min with an apparent rate constant value of 10.3 and 6.4 times higher than the value obtained using ultrasound-H_2O_2 or H_2O_2-Fe_3O_4 system, respectively. This strong synergistic effect might be due to an increase in nucleation sites provided by the Fe_3O_4-MNP which favored the for-mation of cavities and increased the ultrasonic efficiency. The sonochemical effect also enhanced the mass transfer on the surface of Fe_3O_4-MNP, leading to an increase in collision proportionality of the reactants. No observable influence on RhB degradation was found by replacing the catalyst with TiO_2 (Degussa P25) or SiO_2 nanoparticles (Wang et al. 2010b). However, Pang et al. (2011b) showed that sonocatalytic degradation of RhB was effective by using TiO_2 nanotubes synthe-sized using hydrothermal method in the presence of H_2O_2. With an addition of air flow into the system, the degradation of RhB increased from 76 to 85 %. The presence of dissolved oxygen played an important role in generation of OH· while the nitrogen in air might trigger a series of reactions, leading to formation of other reactive species such as NO_2 and NO_3· (Pang et al. 2011b).

Basic Blue 9 (Methylene blue, MB) is another basic dye, which is often used in the textile industry. Thus, MB has also been studied by many research groups (González and Martínez 2008). For examples, Shimizu et al. (2007) carried out a study on sonocatalytic degradation of MB using TiO_2 pellets. They showed that Al_2O_3 pellets had no effect on MB degradation as opposed to TiO_2 pellets (~ 85 % degradation) (Shimizu et al. 2007). Yuan et al. (2009) prepared TiO_2 nanotube array (by using aniodic oxidation method), which was used in sonophotocatalytic degradation of MB. In their study, ultrasonic was found to improve the mass-transfer coefficient by creating high speed microscopic turbulence at solid–liquid interface in porous structure. The use of this recyclable catalyst showed high sonophotocatalytic activity at 27 kHz with synergy as high as 22.1 % (Yuan et al. 2009). Other related studies using MB as a model dye contaminant include son-ocatalytic treatment via TiO_2-CNT (Zhang et al. 2011b) and fullerene/TiO_2 cat-alyst (Meng and Oh 2011).

Li et al. (2010a) conducted comparative experiments to demonstrate the effect of ultrasonic irradiation applied in the electro-Fenton process for Cationic Red X-GRL (Basic Red 46) treatment. When activated carbon fiber was used as cathode, several layers of compounds such as dye molecule, ferrous iron, and ferric complexes were adsorbed, preventing electroreduction from dissolved O_2 as well as H_2O_2 and ferric reduction reaction from occurring. It was found that ultrasonic irradiation helped activate the cathode by providing cavitation near the electrode surface. The continuous cleaning effect provided good electrocatalytic activity for dye degradation. To evaluate the various current efficiencies resulting from the ultrasonic irradiation in the sonoelectro-Fenton process, Li et al. (2010a) also calculated the mineralization current efficiency (MCE) of Cationic Red-XGRL. MCE was significantly enhanced in the presence of ultrasonic irradiation from 43.92 % in the electro-Fenton process to 56.2 % in the sonoelectro-Fenton process with ultrasound power of 160 W (Li et al. 2010a).

3.5.3 Acid Dyes

Acid dyes are water-soluble dyes which are often applied to nylon, wool, silk, and modified acrylics (Hunger 2003). Among all the acid dyes, Acid Orange 7 (AO7) is often subjected to treatment study due to its wide application in the textile and paper industries and AO7 is characterized by its poor biodegradability (Zhong et al. 2011a). Zhang et al. (2009b) showed that ultrasound alone could not decolorize AO7 due to the dye's hydrophilic characteristics and the negligible generation of OH· in bulk liquid solution. By conducting experiments on the effect of ultrasound power density in combined treatment of ultrasound advanced and Fenton process to treat AO7, they concluded that ultrasonic irradiation led to a faster dissolution of iron due to removal or destruction of passivation films on metal surface by cavitation effects and an increase in mass transfer caused by microjetting. However, the difference of final decolorization efficiency using different power densities was not significant, indicating that ultrasonic irradiation only affected the rate of OH· generation and not its amount (Zhang et al. 2009b). Decolorization of AO7 by using ultrasound-geothite-H_2O_2 (Ultrasound enhanced heterogeneous Fenton-like process) system was also investigated (Zhang et al. 2009a). Negligible decolorization of the dye was observed when ultrasound, goethite, or H_2O_2 was used alone. With the presence of ultrasound, heterogeneous Fenton process was enhanced with the dissociating action of the formed $Fe(OOH)^{2+}$ complex on the catalyst surface into Fe^{2+} and HO_2· radicals by sonication. Although high decolorization efficiency was achieved (~ 90 %), the removal of TOC was only 42 % after 90 min, indicating that the difficulty in oxidizing the intermediate products such as carboxyl acids as compared to their parent compound (Zhang et al. 2009a). Other catalysts used in conjunction with ultrasound include Fe-containing mesostructured silica material or Fe_2O_3-Al_2O_3-

meso (Zhong et al. 2011a) and mesoporous alumina supported nanosized Fe_2O_3 (Zhong et al. 2011b).

Zhang et al. (2008a) studied the combination of ultrasound and ozonation in a rectangular air-lift reactor for AO7 degradation. The presence of ultrasonic irradiation enhanced mass transfer rate of ozone from gas phase to liquid phase. When the treatment was conducted without cooling water into the reactor (the system thus rose from 29 to 61 °C), significant improvement of AO7 mineralization was observed (in which case TOC degradation was found to increase from about 40 % to almost 100 %). The elevated temperature would also improve mass transfer of ozone into the solution and enhance the decomposition of ozone to free radicals (Zhang et al. 2008a). The degradation of AO7 using Fe^0/Granular activated carbon (GAC) system under ultrasonic irradiation was investigated by Liu et al. (2007). Synergistic effect was found due to the improvement of overall mass transport and cleaning action of iron chippings which led to Fe^0 activation by ultrasonic irradiation (Liu et al. 2007). Similar improvement was also investigated for the degradation of Acid Red 14 using cast iron in the presence of low frequency ultrasound, where more reactant surface area was formed by the cleaning of cast iron through ultrasonic cavitations (Lin et al. 2008). For the first time, Wang et al. (2008g) reported the use of Au/TiO_2 as a sonocatalyst in the degradation of AO7 (Orange II). Almost no dye degradation was observed using sonication alone. The dye was resistant to sonication because of its non-volatile nature and is highly soluble in water. With the help of Au/TiO_2 under Argon gas atmosphere, sonication rapidly decolorized the dye (100 % in 180 min) and also reduced TOC up to ~ 80 % in 9 h. From ion chromatography, both oxidative and reductive degradation intermediates have been detected, showing that AO7 underwent both reduction by H·and oxidation by OH·produced from the process (Wang et al. 2008g).

For the degradation of azo dye Acid Red B (ARB), Wang et al. (2007a, 2008b, 2009a) studied different kinds of catalyst in combination with ultrasound. Different decolorization efficiency was observed when nanometer anatase and rutile TiO_2 were used in sonocatalytic degradation of ARB. Nanometer anatase TiO_2 showed better performance as the dye was mainly oxidized by the holes on the surface of the particle. By using nanometer rutile TiO_2, ARB molecules were mainly degraded by OH· resulted from the TiO_2 under ultrasonic irradiation (Wang et al. 2007a). Sonocatalytic degradation of ARB using nano-sized ZnO powder assisted by different types of inorganic oxidants ($KClO_4$, $KClO_3$, $Ca(ClO)_2$) were also investigated (Wang et al. 2009a). An addition of oxidants greatly enhanced the degradation as they could react with electrons in conduction band of semiconductor oxides, which resulted in reduction of recombination chance of the electron–hole pair. The existence of these heterogeneous oxides encouraged a large number of active sites appeared on the surface of nano-sized ZnO particles, which then could promote the decomposition of various oxidants under ultrasonic irradiation and generation of added OH· (Wang et al. 2009a).

Ghodbane and Hamdaoui (2009) introduced the use of ultrasonic wave with high frequency (1,700 kHz) on sonochemical decolorization of anthraquinonic dye

Acid Blue 25 (AB25). As compared to ultrasonic irradiation with frequency of 22.5 kHz, treatment using high frequency showed better dye degradation because of the increase in number of acoustic cycles and number of cavitation collapses. Significant intensification of AB25 decolorization in the presence of CCl_4 was also reported. The investigated dosimeter methods, namely KI oxidation, Fricke reaction, and H_2O_2 production, were well corroborated with the improvement of the sonochemical effects in the presence of CCl_4 (Ghodbane and Hamdaoui 2009). Madhavan et al. (2010a) conducted experiments of sonolysis, heterogeneous photocatalysis and their combination on Orange-G degradation. The results showed that even though sonophotocatalysis was the most efficient treatment process, synergistic effect of sonolysis and photocatalysis were not obtained because only a simple additive effect was observed. Aniline, phenol, and aromatic hydroxyl amine were identified as the reaction by-products accompanying Orange-G degradation using high-performance liquid chromatograph (HPLC) and electrospray mass spectrometry (ESMS) (Madhavan et al. 2010a).

3.5.4 Indicator Dyes

Dyes which are usually used as indicators can be characterized by its ability to show different optical properties when the system, of which it forms a part, changes its status and the changes are reversible (Hunger 2003). One of the examples of indicator dyes is Methyl Orange (MO) because it can be reduced or oxidized under different experimental conditions (Bejarano-Pérez and Suárez-Herrera 2007). Complete degradation of MO is necessary as this dye is known to be mutagenic and carcinogenic (Cui et al. 2011). The combination of sonolysis and photocatalytic treatment of MO has been extensively studied in recent years. For example, Bejarano-Pérez and Suárez-Herrera (2007) showed a remarkable increase in oxidation rate of MO using photocatalytic in the presence of ultrasound. They suggested that electric interaction between the bubbles and TiO_2 particles played an important role on the synergistic effect between ultrasound and UV light. The strong electric fields produced by negatively charged microbubbles close to the locally charged TiO_2 particles could induce local discharges that may promote many chemical and physical processes which increase the reaction rate of any reaction on TiO_2 surface. The positive effect of ultrasound was also shown for the reduction process, demonstrated by the photocatalytic reduction process of MO in the presence of ascorbic acid as a hole scavenger. Cui et al. (2011) showed that this synergistic effect only occurred under acidic condition (pH 4). The acidic condition enabled the TiO_2 to be charged positively, while MO existed as the quinone structure was charged negatively, which benefited the adsorption of MO onto the surface of TiO_2. No obvious synergistic effect observed for sonophotocatalysis using TiO_2 for MO degradation at neutral pH (Cui et al. 2011).

Wang et al. (2008a) demonstrated a complete decomposition of MO within 120 min using Ag/TiO_2 illuminated by Xenon lamp coupled with ultrasound treatment. With an addition of radical scavenger such as mannitol or dimethyl sulfoxide (DMSO), although inhibition of MO degradation was observed initially, prolonged reaction time showed a remarkable increase in degradation ratio. UV–vis spectra of MO revealed the presence of intermediates produced by the radical scavenger might be the cause of the acceleration of the degradation (Wang et al. 2008a). On the other hand, Wang et al. (2009b) showed that the addition of ultrasound into photocatalysis process in MO treatment using the prepared CNT/ TiO$_2$ yielded 20 % higher efficiency as compared to photocatalysis alone. Zhang et al. (2008b) took a step further by introducing a highly ordered TiO$_2$ nanotube array, which was fabricated on pure TiO$_2$ sheet with anodization technology in hydrofluoric acid solution to be used as photoelectrode in sonophotoelectrocatalytic process of MO degradation. The rate constant obtained was greater as compared to photoelectrocatalytic or sonophotocatalytic process, suggesting a synergistic effect among photo-, electro-, and sono- processes. In this process, ultrasound provided more OH· for oxidation while the recombination of photogenerated hole/electron pairs was suppressed by external electric field, thus prolonging its lifetime (Zhang et al. 2008b).

Recently, Zhao et al. (2011) reported that the use of zirconia nanotubes can increase the efficiency of ultrasonic treatment of MO by a factor more than 7 times. Zirconia nanotubes were able to adsorb MO molecules in solution, resulting in easy degradation of MO through ultrasonic wave (Zhao et al. 2011). He et al. (2011) investigated the potential of three-dimensional electrode reactor, coupled with ultrasonic in the treatment of wastewater containing MO. Besides producing OH· for dye decomposition, ultrasound had a physical effect whereby the cavitation collapse produced liquid jets to clean the surface of electrode which improved the mass transfer from solid to liquid (He et al. 2011).

Another example of indicator dye is Malachite Green (MG), which is a triarylmethane dye, and widely used as a biocide in aquaculture industry as well as in silk, wool, cotton, leather, paper, and acrylic industries as a dye. Removal of MG from wastewater before discharging is necessary as this chemical is classified as Class II Health Hazard due to its toxicity to human cells (Bejarano-Pérez and Suárez-Herrera 2007, Moumeni and Hamdaoui 2011). Behnajady et al. (2008b) presented a study of ultrasonic degradation of MG using an ultrasonic bath. They showed that the presence of EtOH, PrOH, and iso-BuOH lowered the efficiency of the process by acting as OH· scavengers. Hypsochromic shift in UV–vis spectrum indicated that N-demethylation product was one of the intermediates during the sonication process (Behnajady et al. 2008b). In order to enhance the degradation of MG using ultrasonic irradiation, Moumeni and Hamdaoui (2011) showed the positive effect of increasing bromide ions in wastewater. Although Br$_2$·$^-$ radicals, which are formed by the reaction of Br$^-$ and OH·, are generally less reactive as compared to OH·, they could migrate far away from the cavitation bubbles towards the solution bulk for the degradation of MG. Additionally, these radicals undergoes radical–radical recombination at a lesser extent as compared to OH· and could be more available for substrate degradation at both bubble surface and in solution bulk (Moumeni and Hamdaoui 2011). Bejarano-Pérez and Suárez-Herrera

(2008) showed that MG degradation using sonochemical or sonophotocatalytic reaction was increased by about 2.5 times in the presence of CCl_4. CCl_4 inhibited the recombination of H· and OH· to reform water, thus increasing the level of OH· and improving the overall efficiency. However, treatment time should be prolonged to ensure the degradation of harmful intermediates produced from MG and CCl_4 (Bejarano-Pérez and Suárez-Herrera 2008).

As for other indicator dyes, Guzman-Duque et al. (2011) studied the ultrasonic degradation of Basic Violet 3, also known as Crystal Violet (CV), under different experimental conditions. Although CV was completely removed, the amount of TOC remained high as ultrasonic action led to a transformation of CV to organic by-products with low volatility and high hydrophilic character, which migrated to bulk solution. Thus, the by-products were hardly degraded by pyrolysis or OH· attack. However, ultrasonic treatment significantly enhanced the biodegradability of the solution which was important to ensure a more economical biological treatment (Guzman-Duque et al. 2011). Other studies related to ultrasonic treatment of indicator dye are Congo Red (Wang et al. 2007a; Gopinath et al. 2010) and Acid orange 52 (Maezawa et al. 2007) treatments.

3.5.5 Direct Dyes

Direct dyes are water-soluble anionic dyes which have molecules similar in structure to those of acid dyes but are larger (Ingamells 1993; Hunger 2003). Song et al. (2007) showed that the combination of ozonation and sonolysis was a highly effective way to remove Direct Red 23 from wastewater (98 % degradation in 1 min). Ultrasound waves improved the efficiency by enhancing the efficiency of O_3 dissolution and the yield of free radicals via mechanical action (Song et al. 2007). Chen et al. (2011) showed the synergistic effect of sonication combined with zero-valent iron in the degradation of Direct Sky Blue 5B (Direct Blue 15). This effect was mainly due to the increase in OH· concentration from Fenton's reaction. Changes of UV spectra of dye showed a disappearance of both azo and aromatic groups during the degradation using the process. Song and Li (2009) investigated the use of fly ash combined with ultrasound in the presence of H_2O_2 to treat wastewater containing Direct Black 168. Improvement was observed for treatment using ultrasound, causing the fly ash particles to rupture and thus decreasing the particle size and increasing the surface area available for reaction. Symmetric and asymmetric cavitations by using ultrasonic irradiation also accelerated the heterogeneous reactions for dye degradation (Song and Li 2009).

3.5.6 Food Dyes

There have been increasing concerns on the high concentration of dye effluents from food industries which are potentially harmful if discharged into aquatic environment (Pavanelli et al. 2011). Amaranth or Food Red 9 is an example of food dye which has been approved for food coloration (Hunger 2003). Song et al. (2011a, b) synthesized La^{3+} doped TiO_2 and Tb_7O_{12}/TiO_2 catalyst using sol–gel process and hydrolysis-calcination process, respectively, to be used in sonocatalytic degradation process of Amaranth. In their process, ultrasonic cavitation mechanism generated high energy light, which excited the synthesized semiconductor to produce OH· for dye degradation (Song et al. 2011a, b). Pavanelli et al. (2011) investigated the use of zero-valent metals (Fe and Sn) under ultrasonic irradiation in the treatment of different food industry dyes (Brilliant Blue, Amaranth, Sunset Yellow and Red 40). Under optimized treatment condition using Fe, high dye degradation could be obtained but low mineralization of the dye wastewater was observed due to the reductive characteristic of the process (Pavanelli et al. 2011).

3.6 Ultrasonic Treatment of Pharmaceutical Compounds

Of late, pharmaceutical products become a subject of great interest to environmentalists worldwide because pharmaceutical pollutants have been found in surface water. Studies found that these compounds are highly persistent and able to pass through the treatment plants with relatively little or no degradation. Hence, development of suitable treatment technologies are being carried out to remove the pharmaceutical pollutants from the wastewater before they are discharged into the environment.

Estrogen hormone is either produced naturally in the body or is created for pharmaceutical use for humans and animals. This compound has been reported by many countries to be detected in considerable concentration in surface water, soil, and sediment. Many reports documented the adverse effects of estrogen on living organisms such as feminization of fish in aqueous system. Effective treatment is a necessity to prevent potential risk to human health and living organisms. The degradation performance of eight types of estrogen compound under ultrasound-induced treatment was assessed by manipulating operational conditions (Fu et al. 2007). Low solution pH was found to be more favorable for degradation as more estrogens would exist in nonionic molecular form and show larger hydrophobicity, which eventually caused the estrogens to diffuse easily into the cavity-liquid interface region and undergo both thermal degradation and concentrated radical oxidation upon cavity implosion. Hence, decrease in solution pH led to an increase in reaction rate. Suri et al. (2007) evaluated degradation efficiency of seven estrogens with three different ultrasound systems (0.6, 2, and 4 kW reactor).

Estrogen rate constant were observed to be similar for the 0.6 and 2 kW batch reactors due to the similar ultrasound intensity, whereas significantly higher rate for the 4 kW ultrasound reactor was obtained due to higher power intensity. This study concluded that estrogen degradation rates increased with an increase in power intensity while the energy efficiency of the reactor was higher at lower power density.

Diclofenac is an anti-inflammatory non-steroidal drug, which has become one of the major pollutants found in aqueous environments due to its widespread use and its resistance to biodegradation. This pollutant is reported to exhibit adverse effects on aquatic organisms. Several studies were performed under ultrasonic-enhanced degradation in single or combined treatment to investigate the degradation efficiency of diclofenac. Diclofenac conversion is enhanced at increased applied ultrasound power densities, acidic conditions, and in the presence of dissolved air or oxygen, while the reaction rate increased with increasing diclofenac initial concentration in the range of 2.5–5 mg/L and remained constant in the range of 40–80 mg/L (Naddeo et al. 2010). Madhavan et al. (2010b) reported that diclofenac degradation rate increased with an increase in diclofenac initial concentration until 0.05 mM and then it leveled off. This phenomenon could be explained by considering the formation of OH· by the sonolysis of water. The degradation of diclofenac using TiO_2 and Fe-ZnO was also studied and the degradation rate was observed to be reduced with an addition of TiO_2 or Fe-ZnO. This was due to the scattering of the acoustic waves by the photocatalysts, leading to a low cavitation activity (Madhavan et al. 2010b). However, Hartmann et al. (2008) reported that the addition of catalyst (TiO_2) could enhance the diclofenac degradation because the relative concentration of diclofenac decreased from 100 to 16 % within 30 min of sonolysis. Fe-containing additives such as Fenton's reagent (DVI), zero valent iron (ZVI) and paramagnetic iron oxide nanoparticles (NPI) were added to improve the degradation process efficiency (Güyer et al. 2011). Güyer et al. (2011) reported that a threshold concentration for each additive existed, above which the degradation efficiency dropped off. A combined treatment with ultrasound and O_3 was applied and the effect of this combination on diclofenac degradation was compared with single ultrasound treatment (Naddeo et al. 2009). The combined treatment gave higher removal rate and TOC degradation efficiency as compared to a single ultrasound treatment.

The presence of antibiotics in the aquatic environment becomes an environmental issue due to the potential risk for the emergence or persistence of antibiotic resistance. A degradation of dciprofloxacin (whcih is a non-biodegradable fluoroquinolone antibiotic) was studied by Bel et al. (2009, 2011) under ultrasound treatment by manipulating operational parameters such as pH, frequency, temperature, initial concentration, and others. The studies showed that pH played an important role in determining the degradation rate as the degradation rate constant increased almost fourfold from pH 7 (0.0058 min^{-1}) to pH 3 (0.021 min^{-1}). This phenomenon was attributed to the degree of protonation of ciproflaxin molecule (Bel et al. 2009). Another study proved that degradation constant strongly dependent on the temperature of the bulk solution as the constant increased significantly with increasing temperature (Bel et al. 2011).

Levofloxacin is a synthetic chemotherapeutic antibiotic, which can inhibit microorganisms. This compound is extremely resistant to biological degradation processes and usually escapes intact from conventional treatment plants (Guo et al. 2010a). The effect of operating conditions on the decomposition of levofloxacin under ultrasound irradiation was examined by Guo et al. (2010b). An increase of ultrasound power had a positive effect on the removal rate of COD and a maximum removal rate was observed with ultrasound power of 400 W. This study indicated that the removal rate of COD was effective under a weak acid condition while the removal rate of COD was low under strong acidic or basic conditions. This phenomenon could be explained by considering the characteristic of levofloxacin, which exists as a zwitterion, in which the molecule contains both a basic and acidic group with two different acid dissociation constant values (5.7 and 7.9). The addition of carbon tetrachloride enhanced the degradation significantly as the removal percentage increased from 9.4 to 94.8 % after 20 min of ultrasonic irradiation in the presence of 0.02 mL CCl_4. This improvement could be attributed to the oxidizing species formed during sonolysis of CCl_4, and then to OH· (Guo et al. 2010b). The degradation of levofloxacin was observed to be effectively quenched by an addition of t-butanol in the absence or presence of CCl_4 as t-butanol were able to scavenge OH· in the bubble and reduce the degradation rate significantly (Guo et al. 2010a).

Tetracycline hydrochloride is a well-known class of antibiotics and is consumed globally to resist against infectious diseases for human and veterinary treatment (Wang et al. 2011d). Effect of operational parameters on the degradation of tetracycline was performed in a rectangular air-lift reactor with combined treatments, namely ozonation and ultrasound (Wang et al. 2012). The decomposition rate was observed to be enhanced with an increase of ozone concentration, gas flow rate and power density but decreased with an increase of initial tetracycline concentration. By increasing power density, gas flow rate, and ozone concentration, the mass transfer rate of ozone from gas phase to liquid phase was also increased; hence higher degradation rate was observed. Wang et al. (2011d) also conducted tetracycline degradation study by combining ultrasound, ozone, and goethite catalyst. This study indicated that the degradation performance was enhanced by an addition of goethite as the highest rate constant (0.764/min) was obtained under ultrasound/O_3/goethite combined treatment.

Isariebel et al. (2009) evaluated the influence of operational conditions of the ultrasound treatment on the decomposition of paracetamol and levodapa. They reported that the degradation efficiency and COD removal decreased with an increase of initial solute concentration and a decrease of power. Although increasing in ultrasound frequency could promote the oxidation of both pollutants in relatively dilute synthetic solutions, the extent of degradation strongly depended on the operating conditions. The degradation performance could be enhanced by an addition of H_2O_2 because it could be decomposed by ultrasound to form reactive OH·. However, it could also act as a radical scavenger, depending on the product and the conditions used. Table 3.6 summarizes the degradation performance of pharmaceutical compounds under ultrasonic-enhanced degradation in single or combined treatment.

3.7 Ultrasonic Treatment of Pesticides

The use of pesticides is necessary in order to maintain world food production. It is estimated that losses to pests would increase by 10 % if no pesticides is used and for specific crops, losses could be in the range of zero to nearly 100 %. However, it has also been estimated that only 0.1 % of the applied pesticides reaches the target pest, while 99.9 % is dispersed to the environment (Hart and Pimentel 2002). The widespread of pesticides in water bodies such as surface water, groundwater, and drinking water has been a great concern as it represents a potential threat to both humans and aquatic organisms (Lopes et al. 2008; Zhang et al. 2011c). It should be noted that the combined effects of multi-pesticides may show higher adverse effects on human health, making its removal in drinking water production essential (Zhang et al. 2011c). Table 3.7 shows the recent studies on ultrasound treatment of pesticides.

One of the commonly used agricultural pesticide is DDT (1,1,1-trichloro-2,2-bis(p-chlorophenyl)ethane), which was also widely used to control malarial mosquito populations until the late 1960s. DDT is not only carcinogenic, but also affects the human nervous system and can be transmitted through generations. Despite the global ban on DDT imposed by the 2004 Stockholm Convention, DDT is still been used in developing countries to control malaria under the supervision of United Nations (Thangavadivel et al. 2009). One of the existing methods to dechlorinate DDT is by using microbial transformation pathway. However, it does not satisfactorily address the remediation of DDT due to its resistance to biological degradative reactions and currently, no known microbes have yet to evolve DDT to be used as carbon and energy sources or to completely mineralise it (Gautam and Suresh 2007). Thangavadivel et al. (2009) showed that the use of high frequency ultrasound was effective in degrading the non-polar DDT pollutant, which was dispersed in water and sand slurry. Even though mass transfer amount per bubble was very small due to high oscillation with very short live bubble at high frequency, DDT mass transfer into the cavities during rectified diffusion was significant at any specified time, facilitating its degradation. Addition of low concentration of iron was also found to be effective in increasing the rate of DDT degradation. Thangavadivel et al. (2009) noted that further investigations were required because practically, high frequency transducers are generally thin and fragile, thus are not suitable to be used in heavy duty application such as soil slurry remediation.

In another study, Zouaghi et al. (2011) investigated the degradation of phenylurea monolinuron (MLN), which is a type of herbicide of phenylurea family, by sonolysis and sonocatalysis. Phenylureas are persistent herbicides with high rates

Table 3.6 Degradation performance of pharmaceutical compounds under ultrasonic-enhanced degradation in single/combined treatment

Pharmaceutical compound(s)	Treatment(s)	Initial concentration	Experimental conditions	Results	Reference
Ciprofloxacin	Ultrasound	15 mg/L	520 kHz, 92 W/l, 25 °C, pH 3, 120 min	Degradation rate: 2.10×10^{-2} min^{-1} BOD/COD ratio: 0.60	Bel et al. (2009)
Ciprofloxacin	Ultrasound	15 mg/L	544 kHz glass water-jacketed reactor, 25 °C, pH 7	Degradation efficiency: 50 % Degradation rate: 6.70×10^{-3} min^{-1}	Bel et al. (2011)
Diclofenac	Ultrasound	30 μM (8.88 mg/L)	861 kHz plate-type transducer, 0.23 W/ml, 500 ml glass reactor, 25 °C, 90 min	Degradation rate: 5.21×10^{-2} min^{-1}	Güyer and Ince (2011)
Diclofenac	Ultrasound + Catalyst (ZVI)	30 μM (8.88 mg/L)	861 kHz plate-type transducer, 0.23 W/ml, 500 ml glass reactor, 25 °C, 90 min 8.90MM ZVI	Degradation rate: 9.19×10^{-2} min^{-1}	Güyer and Ince (2011)
Diclofenac	Ultrasound	50 mg/L	617 kHz ultrasound reactor, 30 W, 25 °C, 30 min	Degradation Efficiency: 72 %	Hartmann et al. (2008)
Diclofenac	Ultrasound + Catalyst (TiO$_2$)	50 mg/L	617 kHz ultrasound reactor, 30 W, 25 °C, 30 min 100 mg/L TiO$_2$	Degradation Efficiency: 84 %	Hartmann et al. (2008)
Diclofenac	Ultrasound	0.07 mM (20.73 mg/L)	213 kHz, 55 mW/ml, 25 °C, 240 min	Degradation rate: 15.2×10^{-7} M min^{-1}	Madhavan et al. (2010b)
Diclofenac	Ultrasound + Catalyst (TiO$_2$)	0.07 mM (20.73 mg/L)	213 kHz, 55 mW/ml, 25 °C, 240 min 1 g/L TiO$_2$	Degradation rate: 11.5×10^{-7} M min^{-1}	Madhavan et al. (2010b)

(continued)

Table 3.6 (continued)

Pharmaceutical compound(s)	Treatment(s)	Initial concentration	Experimental conditions	Results	Reference
Diclofenac	Ultrasound	40 mg/L	20 kHz ultrasonic transducer, 400 W/l, 20 °C, 40 min	TOC degradation rate: 0.106 mg l^{-1} min^{-1}	Naddeo et al. (2009)
Diclofenac	Ultrasound + O_3	40 mg/L	20 kHz ultrasonic transducer, 400 W/l, 20 °C, 40 min 31 g/h O_3	TOC degradation rate: 0.211 mg l^{-1} min^{-1}	Naddeo et al. (2009)
Diclofenac	Ultrasound	40 mg/L	20 kHz ultrasound generator, 100 W/l, 300 ml cylindrical vessel, 24 °C, pH 3, 60 min	Degradation efficiency: 50 %	Naddeo et al. (2010)
Estrogen hormones	Ultrasound	50 µg/L	20 kHz sonication reactor, 2.1 W/ml, 20 °C, pH 3, 25 min	Degradation Efficiency: 90 %	Fu et al. (2007)
Estrogen hormones	Ultrasound	10 µg/L	20 kHz sonicator, 4 kW, 3 l continuous flow reactor, 60 min	Estrone degradation rate constant: 0.1513 min 17 α-dihydroequilin degradation rate constant: 0.2313 min Equilin degradation rate constant: 0.0605 min	Suri et al. (2007)
Levodopa	Ultrasound + H_2O_2	25 mg/L	574 kHz ultrasonic generator, 32 W, 0.5 L cylindrical glass vessel, 20 °C 948 mg/L H_2O_2 dosage	Initial degradation rate: 0.60 mg L^{-1} min^{-1}	Isariebel et al. (2009)
Levofloxacin	Ultrasound	20 mg/L	20 kHz ultrasonic generator, 200 W, 21 °C, pH 5.86, 35 min	Degradation efficiency: 12 %	Guo et al. (2010a)
Levofloxacin	Ultrasound	20 mg/L	20 kHz ultrasonic generator, 200 W, 25 °C, pH 5.86, 35 min	Degradation efficiency: 9.4 %	Guo et al. (2010b)
Levofloxacin	Ultrasound + CCL_4	20 mg/L	20 kHz ultrasonic generator, 200 W, 25 °C, pH 5.86, 35 min 0.02 ml CCL_4	Degradation efficiency: 94.8 %	Guo et al. (2010b)

(continued)

Table 3.6 (continued)

Pharmaceutical compound(s)	Treatment(s)	Initial concentration	Experimental conditions	Results	Reference
Levofloxacin	Ultrasound + CCL$_4$	20 mg/L	20 kHz ultrasonic generator, 200 W, 21 °C, pH 5.86, 35 min 0.02 ml CCL$_4$	Degradation efficiency: 97 %	Guo et al. (2010b)
Paracetamol	Ultrasound + H$_2$O$_2$	25 mg/L	574 kHz ultrasonic generator, 32 W, 0.5 L cylindrical glass vessel, 20 °C 590 mg/L H$_2$O$_2$ dosage	Initial degradation rate: 1.80 mg L^{-1} min^{-1}	Isariebel et al. (2009)
Tetracycline	Ultrasound + O$_3$ + catalyst (goethite)	100 mg/L	20 kHz ultrasonic generator, 85.7 W/l, pH 7, 20 min 13.8 mg/L O$_3$ concentration 0.5 g/L goethite	Degradation rate: 0.764 min^{-1}	Wang et al. (2011d)
Tetracycline	Ultrasound	400 mg/L	20 kHz ultrasonic generator, 142.8 W/l, 25 °C, pH 7, 90 min	COD removal: 10 %	Wang et al. (2012)
Tetracycline	Ultrasound + O$_3$	400 mg/L	20 kHz ultrasonic generator, 142.8 W/l, 25 °C, pH 7, 90 min 45.6 mg/L O$_3$ concentration	COD removal: 91 % Degradation efficiency : 100 % Degradation rate: 0.84 min^{-1}	Wang et al. (2012)

Table 3.7 Degradation performance of pesticides pollutants under ultrasonic-enhanced degradation in single/combined treatment

Pesticide(s)	Treatment (s)	Concentration	Experimental conditions	Results	Reference
Carbofuran	Ultrasound + H_2O_2	20 mg/L	20 kHz, 300 W, pH 3, 25 °C, 30 min, 200 mg/L H_2O_2	Degradation: 44 %, TOC removal: 20 %	Ma et al. (2010)
Carbofuran	Ultrasound + Fe^{2+} + H_2O_2	20 mg/L	20 kHz, 300 W, pH 3, 25 °C, 30 min, 100 mg/L H_2O_2, 20 mg/L Fe^{2+}	Degradation: >99 %, TOC removal: 46 %	Ma et al. (2010)
DDT (41 % DDT, 7.4 % DDD, 2.7 % DDE, 0.8 % DDMU, 48.1 % non soluble solid residue carrier)	Ultrasound	8 mg/L (Liquid), 32.6 mg/L (40 wt % sand slurry)	1.6 MHz, 20 W, 40 mL, 90 min	Degradation: 90 % (liquid), 22 % (sand slurry)	Thangavadivel et al. (2009)
Diazinon	Ultrasound	1,200 ppm	1.7 MHz, 9.5 W, 50 mL, 10 min	Degradation: 70 %	Matouq et al. (2008)
Diazinon, chlorpyrifos	Ultrasound	2 mg/L (Diazinon), 2.5 mg/L (Chlorpyrifos)	20 kHz, 600 W, pH 7, 25 °C, 40 min	Degradation: 38 % (Diazinon), 55 % (Chlorpyrifos)	Zhang et al. (2011c)
Dimethoate	Ultrasound + TiO_2	0.39×10^{-4} M (~8.9 mg/L)	22 kHz, 800 W, 500 mL, 90 min, 0.6 g/L TiO_2	Degradation: 11.87 %	Chen et al. (2007b)
Dimethoate	Ultrasound + UV + TiO_2	0.39×10^{-4} M (~8.9 mg/L)	22 kHz, 800 W, 500 mL, 90 min, 0.6 g/L TiO_2, 15 W UV lamp, 235.5 nm	Degradation: 35.02 %	Chen et al. (2007b)

(continued)

Table 3.7 (continued)

Pesticide(s)	Treatment (s)	Concentration	Experimental conditions	Results	Reference
Dimethoate	Ultrasound + O_3	20 mg/L	40 kHz, 250 W, 4.64 W/cm^2, pH 10, 25 °C, 4 h 0.41 m^3/h	Degradation: 90.8 %	Liu et al. (2008)
Linuron	Ultrasound + Fe(II) + UV	10 mg/L	200 kHz, 100 W, pH 3, 100 mL, 25 °C, 20 min 1.2×10^{-4} M Fe(II), 2 mW/cm^2	Degradation: 100 % TOC removal: 100 % (120 min)	Katsumata et al. (2011)
Monolinuron	Ultrasound	4 mg/L	20 kHz, 200 W/L, pH 5.7	Degradation rate constant: ~ 0.02 min^{-1}	Zouaghi et al. (2011)
Monolinuron	Ultrasound + TiO_2	4 mg/L	20 kHz, 200 W/L, pH 5.7 0.2 g/L	Degradation rate constant: ~ 0.035 min^{-1}	Zouaghi et al. (2011)
Parathion	Ultrasound	20 mg/L	20 kHz, 270 W, pH 2.5, 30 °C, 60 min	Degradation: 10.2 % TOC removal: 5.2 %	Shriwas and Gogate (2011b)
Parathion	Ultrasound + H_2O_2	20 mg/L	20 kHz, 270 W, pH 2.5, 30 °C 200 mg/L H_2O_2	Degradation: 15 % (60 min), 97.4 % (72 h) TOC removal: 7.5 % (60 min), 47.3 % (72 h)	Shriwas and Gogate (2011b)

(continued)

Table 3.7 (continued)

Pesticide(s)	Treatment (s)	Concentration	Experimental conditions	Results	Reference
Parathion	Ultrasound + Fenton reaction	20 mg/L	20 kHz, 270 W, pH 2.5, 30 °C, 60 min 200 mg/L H_2O_2, 600 mg/L $FeSO_4$	Degradation: 96 % TOC removal: 73.7 %	Shriwas and Gogate (2011b)
Parathion	Ultrasound	50 mg/L	40 kHz, 50 W, pH 10, 20 °C, 80 min	Degradation: 16.3 %	Wang et al. (2007d)
Parathion	Ultrasound + TiO_2	50 mg/L	40 kHz, 50 W, pH 10, 20 °C, 80 min 1 g/L TiO_2 (Rutile)	Degradation: 95.6 %	Wang et al. (2007d)
Parathion	Ultrasound	2.9 μM (~0.84 mg/L)	600 kHz, 100 W, 0.69 W/cm^2, pH 7, 30 min	Degradation: 99.7 %	Yao et al. (2010b)
Thiamethoxam, Imidacloprid	Ultrasound + Zn^0	50 mg/L	47.3 kHz, pH 2, 20 mL, 30 min 2.5 g/L Zn^0	Degradation: ~100 %	Lopes et al. (2008)

DDT 1,1,1-trichloro-2,2-bis(*p*-chlorophenyl)ethane, *DDD* 2,2-bis(*p*-chlorophenyl)-1,1-dichloroethylene, *DDMU* 1-chloro-2,2-bis(*p*-chlorophenyl)ethylene

of application as total weed killers (Katsumata et al. 2011). They found that the rate constant of MLN degradation was directly proportional to the rate constant of hydroxide produced, independent of the frequency or specific power used. With an addition of nanometric particles (TiO$_2$ or SiO$_2$) into the mixture, sonochemical reaction rate increased at 20 kHz. At low specific power (40 W/L), sonocatalysis at 20 kHz with an addition of TiO$_2$ reached efficiency similar to that obtained by sonolysis at 800 kHz. An addition of nanoparticulates enhanced the reaction yield by increasing the cavitation bubbles, which produced OH· in water for facilitating the degradation of MLN. However, the effects of adding nanoparticulates were insignificant when sonocatalysis was conducted at 800 kHz. This could be due to the added particles were not used as nuclei for generation of cavitation bubbles at high frequency (Zouaghi et al. 2011). Katsumata et al. (2011) demonstrated that the effectiveness of combining ultrasound and photo-Fenton reaction to treat linuron. Complete degradation and mineralization of 10 mg/L linuron could be achieved in 20 min and 120 min, respectively, showing that this combined treatment could be a potentially useful technology for treating linuron (Katsumata et al. 2011).

Diazinon is a form of organophosphorus pesticide, which is often used to control pest insects in soil, ornamental plants, and various field crops. Wastewater from industries which produces diazinon are often contaminated with this pesticide due to the daily cleaning process of equipments and vessels used to synthesize the compound. Frequent discharge of wastewater with high concentration of diazinon is significantly impacting the environments especially on the water resources (Matouq et al. 2008). Matouq et al. (2008) introduced high frequency ultrasound to treat diazinon-contaminated wastewater. With constant ultrasound condition, a degradation of diazinon decreased with increasing volume of wastewater. This process only required 10 min to successfully achieve 70 % diazinon degradation with initial concentration of 1,200 ppm in 50 mL solution volume. The degradation of diazinon was also found to follow a pseudo-first-order model with apparent rate constant of around 0.01 s^{-1} (Matouq et al. 2008). The degradation of diazionon and chlorpyrifos using ultrasonic irradiation was also been investigated by Zhang et al. (2011c). Results showed that these two organophosphorus pesticides could be effectively degraded by ultrasonic irradiation with the extent of degradation, which was strongly dependent on ultrasonic power, temperature, and pH. When mineralization study was performed, the sonication of both organophosphorus pesticides for 60 min did not significantly reduce the TOC of the solution. Based on GC–MS analysis, two and seven degradation products for chlorpyrifos and diazinon were detected, respectively. For chlorpyrifos treatment, the degradation of the compound by using sonication for 60 min was increased by 1.7 times as compared to the at 30 min. However, the toxicity of the sample was increased by 1.1 times, indicating that the high toxicity of the degraded products (chlorpyrifos oxon and 3,5,6-trichloro-2-2pyridinol) as compared to parent compound. On the other hand, toxicity of the sonicated diazinon sample greatly declined by 9.6 times through increasing treatment time from 30 to 60 min (Zhang et al. 2011c).

Another commonly used pesticide is parathion, which is an organophosphorus compound with one structural isomer (Shriwas and Gogate 2011b). For sonochemical treatment of parathion, Yao et al. (2010a) concluded that the free radical reactions were responsible for the degradation with the reaction zones, predominately occurring at the interfacial regions between bubbles and bulk solution and to a lesser extent in bulk solution. Generally, the gas/liquid interfacial regions are effective reaction sites and the reaction can be well described as a gas/liquid heterogeneous reaction, which obeys a kinetic model based on Langmuir–Hinshelwood model (Yao et al. 2010a). The effect of irradiation intensity, dissolved gases, anions, and natural organics on parathion sonolysis was investigated by Yao et al. (2010b). Air bubbling of parathion-containing solution enhanced the sonolysis efficiency as dissolved oxygen greatly improved the degradation. The addition of nitrogen gas into the system slowed down the reactions due to its free radical scavenging effect in vapor phase within the cavitational bubbles. Moreover, unlike Br^- which enhanced the parathion degradation, the presence of CO_3^{2-}, HCO_3^- and Cl^- were found to have inhibiting effects, implying that removal of alkalinity prior to sonolysis could potentially increase the process efficiency. The presence of hydrophobic and hydrophilic natural organic matters also inhibited the parathion degradation rate, particularly the strong hydrophobic component. Since parathion is also a strong hydrophobic organic compound itself, the presence of compounds with similar hydrophobic properties may create competition for reactive oxygen species in the system, especially at the interface region of the cavitational bubbles (Yao et al. 2010b). When investigating the effect of H_2O_2 on the sonolysis of parathion, Shriwas and Gogate (2011b) found that the optimum loading of 10:1 ratio of H_2O_2 to parathion was optimum, giving a removal rate constant of 2.51×10^{-3} min^{-1} (about 15 % removal of 20 ppm parathion). Interesting observations were reported when the samples were kept stand-alone after ultrasonic irradiation for 120 min. They observed an increase in degradation to 70 % in 48 h of treatment time. Without sonication, this observation did not occur for the treatment with H_2O_2. Possible series of chain reaction might occur between the generated radicals and H_2O_2 remaining in the reactor, leading to gradual attack on the pollutants. Besides, the extent of TOC reduction also increased, showing that both parathion and its intermediates were degraded during the extended treatment period. The extent of degradation of parathion by sonolysis using ultrasonic bath or ultrasonic horn was also been compared by the authors. Even though power dissipation levels were significantly higher for ultrasonic horn as compared to ultrasonic bath, similar level of increase in extent of degradation was not observed. This might be due to the more uniform cavitation activity in ultrasonic bath reactor, owing to larger area of transducers for ultrasonic bath (Shriwas and Gogate 2011b).

Ma et al. (2010) studied the degradation of carbofuran, which is a well-known methylcarbamate pesticide, by using ultrasound and Fenton processes. An increase dosage of H_2O_2 and Fe^{2+} would enhance carbofuran degradation in the combined ultrasound-Fenton treatment but an increase in initial concentration reduced the degradation. Kinetic study showed that the degradation of carbofuran followed the first-order kinetics model (Ma et al. 2010). Lopes et al. (2008) investigated the degradation of thiamethoxam and imidacloprid, both are belonged to the neonicotinoid insecticides class, by using various zero-valent metals (Fe, Sn, Zn) exposed to ultrasonic irradiation in water. Ultrasonic irradiation strongly enhanced the degradation rate of both insecticides when zinc and tin were employed. However, the effect of ultrasonic irradiation on the reactivity of zero-valent iron was less pronounced because Fe^0 alone was able to degrade the insecticides. This was probably due to the much lesser tendency of the surface of iron to be passivated, especially in acidic solutions, than tin and zinc. On the other hand, tin and zinc oxides were prone to form an effective passivation layer on the metallic surface that could be efficiently removed upon exposure to ultrasonic irradiation. Lopes et al. (2008) also demonstrated the advantages of using direct-infusion ESI-MSI for the detection and characterization of primary products of reductive degradation of the two insectivides. This was because this technique allowed them to propose a degradation route based on plausible reaction mechanism for the combined treatment (Lopes et al. 2008).

3.8 Conclusion

Ultrasound method is proven to be a promising technology for the degradation of a variety of components in aqueous solution. Although the initial rate of sonochemical degradation is usually high, a complete mineralization is often not achieved in most cases. Oxidation efficiency of OH·is limited by the rates of their generation and their life spans after generation. Instead of increasing the energy input into the process which is economically not favorable, many researchers opted for coupling this technique with other AOPs to provide higher efficiency. Through coupling of AOPs or addition of different types of additives, together with operation under optimized conditions, concentration of reactive radicals are able to be maintained at high levels and thus, effectively degrade the targeted pollutant compound.

References

Aarthi T, Shaama MS, Madras G (2007) Degradation of water soluble polymers under combined ultrasonic and ultraviolet radiation. Ind Eng Chem Res 46:5204–6210

Abbasi M, Asl NR (2008) Sonochemical degradation of Basic Blue 41 dye assisted by nanoTiO$_2$ and H$_2$O$_2$. J Hazard Mater 153:942–947

Akyüz A, Catalgil-Giz H, Giz AT (2008) Kinetics of ultrasonic polymer degradation: comparison of theoretical models with on-line data. Macromol Chem Phys 209:801–809

Ayyildiz O, Peters RW, Anderson PR (2007) Sonolytic degradation of halogenated organic compounds in groundwater: mass transfer effects. Ultrason Sonochem 14:163–172

Behnajady MA, Modirshahla N, Shokri M, Vahid B (2008a) Effect of operational parameters on degradation of malachite green by ultrasonic irradiation. Ultrason Sonochem 15:1009–1014

Behnajady MA, Modirshahla N, Shokri M, Vahid B (2008b) Investigation of the effect of ultrasonic waves on the enhancement of efficiency of direct photolysis and photooxidation processes on the removal of a model contaminant from textile industry. Global Nest J 10:8–15

Behnajady MA, Modirshahla N, Tabrizi SB, Molanee S (2008c) Ultrasonic degradation of Rhodamine B in aqueous solution: influence of operational parameters. J Hazard Mater 152:381–386

Bejarano-Pérez NJ, Suárez-Herrera MF (2007) Sonophotocatalytic degradation of congo red and methyl orange in the presence of TiO_2 as a catalyst. Ultrason Sonochem 14:589–595

Bejarano-Pérez NJ, Suárez-Herrera MF (2008) Sonochemical and sonophotocatalytic degradation of malachite green: the effect of carbon tetrachloride on reactive rates. Ultrason Sonochem 15:612–617

Bel ED, Dequlf J, Witte BD, Langenhove HV, Janssen C (2009) Influence of pH on the sonolysis of ciprofloxacin: Biodegradability, ecotoxicity and antibiotic activity of its degradation products. Chemosphere 77:291–295

Bel ED, Janssen C, Smet SD, Langenhove HV, Dewulf J (2011) Sonolysis of ciprofloxacin in aqueous solution: Influence of operational parameters. Ultrason Sonochem 18:184–189

Bremner DH, Di Carlo S, Chakinala AG, Cravotto G (2008) Mineralisation of 2,4-dichlorophenoxyacetic acid by acoustic or hydrodynamic cavitation in conjunction with the advanced Fenton process. Ultrason Sonochem 15:416–419

Bremner DH, Molina R, Martinez F, Melero JA, Segura Y (2009) Degradation of phenolinc aqueous solutions by high frequency sono-Fenton systems (US-Fe_2O_3/SBA-15-H_2O_2). Appl Catal B-Environ 90(380):388

Chan SHS, Wu TY, Juan JC, Teh CY (2011) Recent developments of metal oxide semiconductors as photocatalysts in advanced oxidation processes (AOPs) for treatment of dye waste-water. J Chem Technol Biotechnol 86:1130–1158

Chand R, Ince NH, Gogate PR, Bremner DH (2009) Phenol degradation using 20, 300 and 520 kHz ultrasonic reactors with hydrogen peroxide, ozone and zero valent metals. Sep Purif Technol 67:103–109

Chang JH, Ellis AV, Yan CT, Tung CH (2009) The electrochemical phenomena and kinetics of EDTA-copper wastewater reclamation by electrodeposition and ultrasound. Sep Purif Technol 68:216–221

Chang CY, Chang YJ, Hsieh YH, Lin CH, Yen SH (2010) Azo dye-Yellow 17 wastewater photocatalytic degradation of by UV/TiO_2 combined with ultrasonic procedure. Adv Mater Res 123–125:11–14

Chen WS, Huang GC (2009) Sonochemical decomposition of dinitrotoluenes and trinitrotulene in wastewater. J Hazard Mater 169:868–874

Chen J, Li H, Lai SY, Jow J (2007a) Degradation and in situ reaction of polyolefin elastomers in the melt state induced by ultrasonic irradiation. J Appl Polym Sci 106:138–145

Chen JQ, Wang D, Zhu MX, Gao CJ (2007b) Photocatalytic degradation of dimethoate using nanosized TiO_2 powder. Desalination 207:87–94

Chen CQ, Cheng JP, Tang Z, Pang C, Zheng ZX (2008) Degradation of azo dyes by hybrid ultrasound-fenton reagent. Bioinformatics and Biomedical Engineering, ICBBE. 2nd international conference

Chen B, Wang X, Wang C, Jiang W, Li S (2011) Degradation of azo dye direct sky blue 5B by sonication combined with zero-valent ion. Ultrason Sonochem 18:1091–1096

Chiha M, Hamdaoui O, Baup S, Gondrexon N (2011) Sonolytic degradation of endocrine disrupting chemical 4-cumylphenol in water. Ultrason Sonochem 18:943–950

Cui P, Chen Y, Chen G (2011) Degradation of low concentration methyl orange in aqueous solution through sonophotocatalysis with simultaneous recovery of photocatalyst by ceramic membrane microfiltration. Ind Eng Chem Res 50:3947–3954

Daraboina N, Madras G (2009) Kinetics of the ultrasonic degradation of poly (alkyl methacrylates). Ultrason Sonochem 16:273–279

Dehghani MD, Mesdaghinia AR, Nasseri S, Mahvi AH, Azam K (2008) Application of SCR technology for degradation of reactive yellow dye in aqueous solution. Water Qual Res J Can 43:1–10

Desai V, Shenoy MA, Gogate PR (2008) Ultrasonic degradation of low-density polyethylene. Chem Eng Process 47:1451–1455

Eren Z, Ince NH (2010) Sonolytic and sonocatalytic degradation of azo dyes by low and high frequency ultrasound. J Hazard Mater 177:1019–1024

Esclapez MD, Sáez V, Milán-Yáñez D, Louisnard O, González-García J (2010) Sonoelectrochemical treatment of water polluted with trichloroacetic acid: from sonovoltammetry to prepilot plant scale. Ultrason Sonochem 17:1010–1020

Fındık S, Gündüz G (2007) Sonolytic degradation of acetic acid in aqueous solutions. Ultrason Sonochem 14:157–162

Fu H, Suri RPS, Chimchirian RF, Helmig E, Constable R (2007) Ultrasound-induced destruction of low levels of estrogen hormones in aqueous solutions. Environ Sci Technol 41:5869–58744

Gao J, Jiang R, Wang J, Kang P, Wang B, Li Y, Li K, Zhang X (2011a) The investigation of sonocatalytic activity of Er^{3+}:$YAlO_3$/TiO_2-ZnO composite in azo dyes degradation. Ultrason Sonochem 18:541–548

Gao J, Jiang R, Wang J, Wang B, Kai L, Kang P, Li Y, Zhang X (2011b) Sonocatalytic performance of Er^{3+}:$YAlO_3$/TiO_2-Fe_2O_3 in organic dye degradation. Chem Eng J 168:1041–1048

Gao Y, Cheng C, Wang J, Wang Z, Jin X, Li K, Kang P, Gao J (2011c) Detection of reactive oxygen species (ROS) generated by TiO_2(R), TiO_2(R/A) and TiO_2(A) under ultrasonic and solar light irradiation and application in degradation of organic dyes. J Hazard Mater 192:786–793

Gautam SK, Suresh S (2007) Studies on dechlorination of DDT (1,1,1-trichloro-2,2-bis(4-chlorophenyl)ethane using magnesium/palladium bimetallic system. J Hazard Mater B139:146–153

Ghodbane H, Hamdaoui O (2009) Intensification of sonochemical decolorization of anthraquinonic dye Acid Blue 25 using carbon tetrachloride. Ultrason Sonochem 16:455–461

González AS, Martínez SS (2008) Study of the sonophotocatalytic degradation of basic blue 9 industrial textile dye over slurry titanium dioxide and influencing factors. Ultrason Sonochem 15:1038–1042

Gopinath KP, Muthukumar K, Velan M (2010) Sonochemical degradation of congo red: optimization through response surface methodology. Chem Eng J 157:427–433

Grčić I, Obradocić M, Vujević D, Koprivanac N (2010a) Sono-Fenton oxidation of formic acid/formate ions in an aqueous solution: from an experimental design to the mechanistic modelling. Chem Eng J 164:196–207

Gronroos A, Pentti P, Hanna K (2008) Ultrasonic degradation of aqueous carboxylmethycellulose: effect of viscosity, molecular mass and concentration. Ultrason Sonochem 15:644–648

Gultekin I, Ince NI (2008) Ultrasonic destruction of bisphenol-A: the operating parameters. Ultrason Sonochem 15:524–529

Guo Z, Feng R (2009) Ultrasonic irradiation-induced degradation of low-concentration bisphenol A in aqueous solution. J Hazard Mater 163:855–860

Guo WY, Peng B (2007) Ultrasonic degradation studies and its effect on thermal properties of polypropylene. Polym-Plast Technol 46:879–884

Guo Z, Feng R, Li J, Zheng Z, Zheng Y (2008) Degradation of 2,4-dinitrophenol by combining sonolysis and different additives. J Hazard Mater 158:164–169

Guo W, Shi Y, Wang H, Yang H, Zhang G (2010a) Intensification of sonochemical degradation of antibiotics levofloxacin using carbon tetrachloride. Ultrasound Sonochem 17:680–684

Guo W, Shi Y, Wang H, Yang H, Zhang G (2010b) Sonochemical decomposition of levofloxacin in aqueous solution. Water Environ Res 82:696–700

Güyer GT, Ince NH (2011) Degradation of diclofenac in water by homogeneous and heterogeneous sonolysis. Ultrason Sonochem 18:114–119

Guzman-Duque F, Pétrier C, Pulgarin C, Peñuela G, Torres-Palma RA (2011) Effects of sonochemical parameters and inorganic ions during the sonochemical degradation of crystal violet in water. Ultrason Sonochem 18:440–446

Hamdaoui O, Naffrechoux E (2008) Sonochemical and photosonochemical degradation of 4-chlorophenol in aqueous media. Ultrason Sonochem 15:981–987

Hart KA, Pimentel D (2002) Environmental and economic costs of pesticide use. In: Pimentel D (ed) Encyclopedia of pest management. CRC Press, Boca Raton

Hartmann J, Bartels P, Mau U, Witter M, Tumpling WV, Hofmann J, Nietzschmann E (2008) Degradation of the drug diclofenac in water by sonolysis in presence of catalyst. Chemosphere 70:453–461

Hayashi N, Yasutomi R, Kasai E (2010) Development of dispersed-type sonophotocatalytic process using piezoelectric effect caused by ultrasonic resonance. Ultrason Sonochem 17:884–891

He Z, Song S, Xia M, Qiu J, Ying H, Lü B, Jiang Y, Chen J (2007a) Mineralization of C.I. Reactive Yellow 84 in aqueous solution by sonolytic ozonation. Chemosphere 69:191–199

He Z, Song S, Ying H, Xu L, Chem J (2007b) p-Aminophenol degradation by ozonation combined with sonolysis: operating conditions influence and mechanism. Ultrason Sonochem 14:568–574

He Z, Lin L, Song S, Xia M, Xu L, Ying H, Chen J (2008) Mineralization of C.I. Reactive Blue 19 by ozonation combined with sonolysis: performance optimization and degradation mechanism. Sep Purif Technol 62:376–381

He P, Wang L, Cao Z (2011) Electrolytic treatment of methyl orange in aqueous solution using three-dimensional electrode reactor coupling ultrasonics. Environ Technol 31:417–422

Hsieh LL, Kang HJ, Shyu HL, Chang CY (2009) Optimal degradation of dye wastewater by ultrasound/Fenton method in the presence of nanoscale iron. Water Sci Technol 60:1295–1301

Hunger K (2003) Industrial dyes: chemistry, properties, application. Wiley, Weinheim

Ingamells W (1993) Colour for textiles: a user's handbook. Society of Dyers and Colourist, West Yorkshire

Inoue M, Masuda Y, Okada F, Sakurai A, Takahashi I, Sakakibara M (2008) Degradation of bisphenol A using sonochemical reactions. Water Res 42:1379–1386

Isariebel QP, Carine JL, Ulises-Javier JH, Anne-Marie W, Henri D (2009) Sonolysis of levodopa and paracetamol in aqueous solutions. Ultrason Sonochem 16:610–616

Isaza PA, Gaugulis AJ (2009) Ultrasonically enhanced delivery and degradation of PAHs in a polymer-liquid partitioning system by a microbial consortium. Biotechnol Bioeng 104:91–101

Jain R, Mathur M, Sikarwar S, Mittal A (2007) Removal of the hazardous dye rhodamine B through photocatalytic and adsorption treatments. J Environ Manage 85:956–964

Jamalluddin NA, Abdullah AZ (2011) Reactive dye degradation by combined Fe(III)/TiO$_2$ catalyst and ultrasonic irradiation: effect of Fe(III) loading and calcination temperature. Ultrason Sonochem 18:669–678

Jiang Mm, Yan Xm, Zou Hs (2009) Coupled methods for effective degradation of harmful chlorobenzene. Bioinformatics and Biomedical Engineering, ICBBE. 3rd international conference

Joseph CG, Puma GL, Bono A, Taufiq-Yap YH, Krishnaiah D (2011) Operationg parameters and synergistic effects of combining ultrasound and ultraviolet irradiation in the degradation of 2,4,6-trichlorophenol. Desalination 276:303–309

Kanwal F, Pethrick RA, Jamil T (2007) Ultrasonic degradation studies of polyacrylates. J Chem Soc Pakistan 29:433–437

Katsumata H, Kaneco S, Suzuki T, Ohta K, Yobiko Y (2007) Sonochemical degradation of 2,3,7,8-tetrachlorodibenzo-p-dioxins in aqueous solution with Fe(III)/UV system. Chemosphere 69:1261–1266

Katsumata H, Kobayashi T, Kaneco S, Suzuki T, Ohta K (2011) Degradation of linuron by ultrasound combined with photo-Fenton treatment. Chem Eng J 166:468–473

Kaur S, Singh V (2007) Visible light induced sonophotocatalytic degradation of Reactive Red dye 198 using dye sensitized TiO_2. Ultrason Sonochem 14:531–537

Kavitha SK, Palanisamy PN (2011) Photocatalytic and sonophotocatalytic degradation of Reactive Red 120 using dye sensitized TiO_2 under visible light. World Acad Sci Eng Technol 73:1–6

Khokhawala IM, Gogate PR (2010) Degradation of phenol using a combination of ultrasonic and UV irradiations at pilot scale operation. Ultrason Sonochem 17:833–838

Khokhawala IM, Gogate PR (2011) Intensification of sonochemical degradation of phenol using additives at pilot scale operation. Water Sci Technol 63:2547–2552

Kidak R, Ince NH (2007) Catalysis of advanced oxidation reactions by ultrasound: a case study with phenol. J Hazard Mater 146:630–635

Kim JK, Martinez F, Metcalfe IS (2007) The beneficial role of use of ultrasound in heterogeneous Fenton-like system over supported copper catalysts for degradation of p-chlorophenol. Catal Today 124:224–231

Koda S, Taguchi K, Futamura K (2011) Effects of frequency and a radical scavenger on ultrasonic degradation of water-soluble polymers. Ultrason Sonochem 18:276–281

Kritikos DE, Xekoukoulotakis NP, Psillakis E, Mantzavinos D (2007) Photocatalytic degradation of reactive black 5 in aqueous solutions: effect of operating conditions and coupling with ultrasound irradiation. Water Res 41:2236–2246

Kubo M, Fukuda H, Chua XJ, Yonemoto T (2007) Kinetics of ultrasonic degradation of phenol in the presence of composite particles of titanium dioxide and activated carbon. Ind Eng Chem Res 46:699–704

Laxmi PNV, Saritha P, Rambabu N, Himabindu V, Anjaneyulu Y (2010) Sonochemical degradation of 2chloro-5methyl phenol assisted by TiO_2 and H_2O_2. J Hazard Mat 174:151–155

Lee M, Oh J (2010) Sonolysis of trichloroethylene and carbon tetrachloride in aqueous solution. Ultrason Sonochem 17:207–212

Li JT, Song YL (2010) Degradation of AR 97 aqueous solution by combination of ultrasound and fenton reagent. Environ Prog Sustain 29:101–106

Li H, Lei H, Yu Q, Li Q, Li Z, Feng X, Yang B (2010a) Effect of low frequency ultrasonic irradiation on the sonoelectro-Fenton degradation of cationic red X-GRL. Chem Eng J 160:417–422

Li JT, Bai B, Song YL (2010b) Degradation of Acid orange 3 in aqueous solution by combination by Fly ash/H_2O_2 and ultrasound irradiation. Indian J Chem Tech 17:198–203

Liang J, Komarov S, Hayashi N, Kasai E (2007) Improvement in sonochemical degradation of 4-chlorophenol by combined use of Fenton-like reagents. Ultrason Sonochem 14:201–207

Lim M, Son Y, Khim J (2011) Frequency effects on the sonochemical degradation of chlorinated compounds. Ultrason Sonochem 18:460–465

Lin JJ, Zhao XS, Liu D, Yu ZG, Zhang Y, Xu H (2008) The decoloration and mineralization of azo dye C.I. Acid Red 14 by sonochemical process: Rate improvement via Fenton's reactions. J Hazard Mater 157:541–546

Liu H, Li G, Qu J, Liu H (2007) Degradation of azo dye Acid Orange 7 in water by Fe^0/granular activated carbon system in the presence of ultrasound. J Hazard Mater 144:180–186

Liu YN, Jin D, Lu XP, Han PF (2008) Study on degradation of dimethoate solution in ultrasonic airlift loop reactor. Ultrason Sonochem 15:755–760

Liu H, Liang MY, Li CS, Gao YX, Zhou JM (2009) Catalytic degradation of phenol in sonolysis by coal ash and H_2O_2/O_3. Chem Eng J 153:131–137

Lopes RP, de Urzedo APFM, Nascentes CC, Augusti R (2008) Degradation of the insecticides thiamethoxam and imidacloprid by zero-valent metals exposed to ultrasonic irradiation in water medium: electrospray ionization mass spectrometry monitoring. Rapid Commun Mass Spectrom 22:3472–3480

Low FCF, Wu TY, Teh CY, Juan JC, Balasubramanian N (2012) Investigation into photocatalytic decolorisation of CI Reactive Black 5 using titanium dioxide nanopowder. Color Technol 128:44–50

Ma YS, Sung CF, Lin JG (2010) Degradation of carbofuran in aqueous solution by ultrasound and Fenton processes: effect of system parameters and kinetic study. J Hazard Mater 178:320–325

Madhavan J, Grieser F, Ashokkumar M (2010a) Degradation of orange-G by advanced oxidation processes. Ultrason Sonochem 17:338–343

Madhavan J, Kumar PSS, Anandan S, Zhou M, Grieser F, Ashokkumar M (2010b) Ultrasound assisted photocatalytic degradation of diclofenac in an aqueous environment. Chemosphere 80:747–752

Maezawa A, Nakadoi H, Suzuki K, Furusawa T, Suzuki Y, Uchida S (2007) Treatment of dye wastewater by using photo-catalytic oxidation with sonification. Ultrason Sonochem 14:615–620

Maleki A, Mahvi AH, Mesdaghinia A, Naddafi K (2007) Degradation and toxicity reduction of phenol by ultrasound waves. Bull Chem Soc Ethiop 21:33–38

Maleki A, Mahvi AH, Ebrahimi R, Zandsalimi Y (2010) Study of photochemical and sonochemical processes efficiency for degradation of dyes in aqueous solution. Korean J Chem Eng 27:1805–1810

Matouq MA, Al-Anber ZA, Tagawa T, Aljbour S, Al-Shannag M (2008) Degradation of dissolved diazinon pesticide in water using the high frequency of ultrasonic wave. Ultrason Sonochem 15:869–874

MdM Rashid, Sato C (2011) Photolysis, sonolysis and photosonolysis of trichloroethane (TCA), trichloroethylene (TCE), and tetrachloroethylene (PCE) without catalyst. Water Air Soil Pollut 216:429–440

Mehrdad A (2011) Ultrasonic degradation of polyvinyl pyrrolidone in mixed water/acetone. J Appl Polym Sci 120:3701–3708

Mehrdad A, Hashemzadeh R (2010) Ultrasonic degradation of Rhodamine B in the presence of hydrogen peroxide and some metal oxide. Ultrason Sonochem 17:168–172

Meng ZD, Oh WC (2011) Sonocatalytic degradation and catalytic activities for MB solution of Fe treated fullerene/TiO_2 composite with different ultrasonic intensity. Ultrason Sonochem 18:757–764

Merouani S, Hamdaoui O, Saoudi F, Chiha M (2010a) Sonochemical degradation of Rhodamine B in aqueous phase: effects of additives. Chem Eng J 158:550–557

Merouani S, Hamdaoui O, Saoudi F, Chiha M, Pétrier C (2010b) Influence of bicarbonate and carbonate ions on sonochemical degradation of Rhodamine B in aqueous phase. J Hazard Mater 175:593–599

Mishra KP, Gogate PR (2011a) Intensification of sonophotocatalytic degradation of p-nitrophenol at pilot scale capacity. Ultrason Sonochem 18:739–744

Mishra KP, Gogate PR (2011b) Intensification of degradation of aqueous solutions of rhodamine B using sonochemical reactors at operating capacity of 7 L. J Environ Manage 92:1972–1977

Moumeni O, Hamdaoui O (2011) Intensification of sonochemical degradation of malachite green by bromine ions. Ultrason Sonochem 19:404–409

Naddeo V, Belgiorno V, Ricco D, Kassinos D (2009) Degradation of diclofenac during sonolysis, ozonation and their simultaneous application. Ultrason Sonochem 16:790–794

Naddeo V, Belgiorno V, Kassinos D, Mantzavinos D, Meric S (2010) Ultrasonic degradation, mineralization and detoxification of diclofenac in water: optimization of operating parameters. Ultrason Sonochem 17:179–185

Nakui H, Okitsu K, Maeda Y, Nishimura R (2007) Effect of coal ash on sonochemical degradation of phenol in water. Ultrason Sonochem 14:191–196

Navarro NM, Chave T, Pochon P, Bisel I, Nikitenko SI (2011) Effect of ultrasonic frequency on the mechanism of formic acid sonolysis. J Phys Chem B 115:2024–2029

Neppolian B, Ciceri L, Bianchi CL, Grieser F, Ashokkumar M (2011) Sonophotocatalytic degradation of 4-chlorophenol using Bi_2O_3/$TiZrO_4$ as a visible light responsive photocatalyst. Ultrason Sonochem 18:135–191

Ostlund SG, Striegel AM (2008) Ultrasonic degradation of poly(γ-benzyl-L-glutamate), an archetypal highly extended polymer. Polym Degrad Stabil 93:1510–1514

Pang YL, Abdullah AZ, Bhatia S (2011a) Optimization of sonocatalytic degradation of rhodamine B in aqueous solution in the presence of TiO_2 nanotubes using response surface methodology. Chem Eng J 166:873–880

Pang YL, Bhatia S, Abdullah AZ (2011b) Process behaviour of TiO_2 nanotube-enhanced sonocatalytic degradation of Rhodamine B in aqueous solution. Sep Purif Technol 77:331–338

Pavanelli SP, Bispo GL, Nascentes CC, Augusti R (2011) Degradation of food dyes by zero-valent metals exposed to ultrasonic irradiation in water medium: optimization and electrospray ionization mass spectrometry monitoring. J Braz Chem Soc 22:111–119

Pilli S, Bhunia P, Yan S, LeBlanc RJ, Tyagi RD (2011) Ultrasonic pretreatment of sludge: a review. Ultrason Sonochem 18:1–18

Pradhan AA, Gogate PR (2010) Degradation of p-nitrophenol using acoustic cavitation and Fenton chemistry. J Hazard Mater 173:517–522

Quan Y, Chen L (2011) Kinetic model of degradation of 2,4-dichlorophenoxyacetic acid in aqueous solution using ultrasound-enhanced ozonation. Remote Sensing, Environment and Transportation Engineering, RSETE. 2011 International Conference

Rokhina EV, Lahtinen M, Nolte MCM, Virkutyte J (2009) The influence of ultrasound on the Rul3-catalyzed oxidation of phenol: catalyst study and experimental design. Appl Catal B-Environ 97:162–170

Sáez V, Esclapez MD, Bonete P, Walton DJ, Rehorek A, Louisnard O, González-García J (2011a) Sonochemical degradation of perchloroethylene: the influence of ultrasonic variables, and the identification of products. Ultrason Sonochem 18:104–113

Sáez V, Esclapez MD, Tudela I, Bonete P, Louisnard O, González-García J (2010) 20 kHz sonoelectrochemical degradation of perchloroethylene in sodium sulphate aqueous media: Influence of the operational variables in batch mode. J Hazard Mater 183:648–654

Sáez V, Tudela I, Esclapez MD, Bonete P, Louisnard O, González-García J (2011b) Sonoelectrochemical degradation of perchloroethylene in water: enhancement of the process by the absence of background electrolyte. Chem Eng J 168:649–655

Segura Y, Molina R, Martinez F, Melero JA (2009) Integrated heterogeneous sono-photo Fenton processes for the degradation of phenolic aqueous solutions. Ultrason Sonochem 16:417–424

Shimizu N, Ogino C, Dadjour MJ, Murata T (2007) Sonocatalytic degradation of methylene blue with TiO_2 pellets in water. Ultrason Sonochem 14:184–190

Shimizu N, Ogino C, Dadjour MF, Ninomiya K, Fujihira A, Sakiyama K (2008) Sonocataytic facilitation of hydroxyl radical generation in the presence of TiO_2. Ultrason Sonochem 15:988–994

Shriwas AK, Gogate PR (2011a) Intensification of degradation of 2,4,6-Trichlorophenol using sonochemical reactors: understanding mechanism and scale-up aspects. Ind Eng Chem Res 50:9601–9608

Shriwas AK, Gogate PR (2011b) Ultrasonic degradation of methyl Parathion in aqueous solution: intensification using additives and scale up aspects. Sep Purif Technol 79:1–7

Song YL, Li JT (2009) Degradation of C.I. Direct Black 168 from aqueous solution by fly ash/H_2O_2 combining ultrasound. Ultrason Sonochem 16:440–444

Song S, Ying H, He Z, Chen J (2007) Mechanism of decolorization and degradation of CI Direct Red 23 by ozonation combined with sonolysis. Chemosphere 66:1782–1788

Song L, Chen C, Zhang S (2011a) Sonocatalytic performance of Tb_7O_{12}/TiO_2 composite under ultrasonic irradiation. Ultrason Sonochem 18:713–717

Song L, Chen C, Zhang S, Wei Q (2011b) Sonocatalytic degradation of amaranth catalysed by La^{3+} doped TiO_2 under ultrasonic irradiation. Ultrason Sonochem 18:1057–1061

Sun JH, Sun SP, Sun JY, Sun RX, Qiao LP, Guo HQ, Fan MH (2007) Degradation of azo dye Acid black 1 using low concentration iron of Fenton process facilitated by ultrasonic irradiation. Ultrason Sonochem 14:761–766

Suri RPS, Nayak M, Devaiah U, Helmig E (2007) Ultrasound assisted destruction of estrogen hormones in aqueous solution: effect of power density, power intensity and reactor configuration. J Hazard Mater 146:472–478

Tauber MM, Gübitz GM, Rehorek A (2008) Degradation of azo dyes by oxidative processes—Laccase and ultrasound treatment. Bioresource Technol 99:4213–4220

Thangavadivel K, Megharaj M, Smart RSC, Lesniewski PJ, Naidu R (2009) Application of high frequency ultrasound in the destruction of DDT in contaminated sand and water. J Hazard Mater 168:1380–1386

Thangavadivel K, Megharaj M, Mudhoo A, Naidu R (2011) Degradation of organic pollutants using ultrasound. In: Chen D, Sharma SK, Mudhoo A (eds) Handbook on applications of ultrasound: sonochemistry for sustainability. CRC Press, Boca Raton

Torres RA, Abdelmalek F, Combet E, Petrier C, Pulgarin C (2007a) A comparative study of ultrasonic cavitation and Fenton's reagent for bisphenol A degradation in deionised and natural waters. J Hazard Mater 146:546–551

Torres RA, Petrier C, Combet E, Moulet F, Pulgarin C (2007b) Bisphenol A mineralization by integrated ultrasound-UV-iron (II) treatment. Environ Sci Technol 41:297–302

Torres RA, Petrier C, Combet E, Carrier M, Pulgarin C (2008) Ultrasonic cavitation applied to the treatment of bisphenol A: effect of sonochemical parameters and analysis of BPA by-products. Ultrason Sonochem 15:605–611

Torres-Palma RA, Nieto JI, Combet E, Petrier C, Pulgarin C (2010) An innovative ultrasound, Fe^{2+} and TiO_2 photoassisted process for bisphenol a mineralization. Water Res 44:2245–2252

Uddin H, Hayashi S (2009) Effects of dissolved gases and pH on sonolysis of 2,4-dichlorophenol. J Hazard Mater 170:1273–1276

Vajnhandl S, Marechal AML (2007) Case study of the sonochemical decolouration of textile azo dye Reactive Black 5. J Hazard Mater 141:329–335

Vecitis CD, Lesko T, Colussi AJ, Hoffmann MR (2010) Sonolytic degradation of aqueous bioxalate in the presence of ozone. J Phys Chem A 114:4968–4980

Vinu R, Madras G (2011) Kinetics of sono-photooxidative degradation of poly(alkyl methacrylate)s. Ultrason Sonochem 18:608–616

Wang J, Jiang Y, Zhang Z, Zhang X, Ma T, Zhang G, Zhao G, Zhang P, Li Y (2007a) Investigation on the sonocatalytic degradation of acid red B in the presence of nanometer TiO_2 catalyst and comparison of catalytic activities of anatase and rutile TiO_2 powders. Ultrason Sonochem 14:545–551

Wang J, Jiang Y, Zhang Z, Zhao G, Zhang G, Ma T, Sun W (2007b) Investigation on the sonocatalytic degradation of congo red catalysed by nanometer rutile TiO_2 powder and various influencing factors. Desalination 216:196–208

Wang J, Pan Z, Zhang Z, Zhang X, Jiang Y, Ma T, Wen F, Li Y, Zhang P (2007c) The investigation on ultrasonic degradation of acid fuchsine in the presence of ordinary and nanometer rutile TiO_2 and the comparison of their sonocatalytic activities. Dyes Pigments 74:525–530

Wang J, Sun W, Zhang Z, Zhang X, Li R, Ma T, Zhang P, Li Y (2007d) Sonocatalytic degradation of methyl parathion in the presence of micron-sized and nano-sized rutile titanium dioxide catalysts and comparison of their sonocatalytic abilities. J Mol Catal A Chem 272:84–90

Wang H, Niu J, Long X, He L (2008a) Sonophotocatalytic degradation of methyl orange by nano-sized Ag/TiO_2 particles in aqueous solutions. Ultrason Sonochem 15:386–392

Wang J, Jiang Z, Zhang Z, Xie Y, Wang X, Xing Z, Xu R, Zhang X (2008b) Sonocatalytic degradation of acid red B and rhodamine B catalysed by nano-sized ZnO powder under ultrasonic irradiation. Ultrason Sonochem 15:768–774

Wang J, Sun W, Zhang Z, Jiang Z, Wang X, Xu R, Li R, Zhang X (2008c) Preparation of Fe-doped mixed crystal TiO_2 catalyst and investigation of its sonocatalytic activity during degradation of azo fuchsine under ultrasonic irradiation. J Colloid Interf Sci 320:202–209

Wang J, Sun W, Zhang Z, Xing Z, Xu R, Li R, Li Y, Zhang X (2008d) Treatment of nano-sized rutile phase TiO_2 powder under ultrasonic irradiation in hydrogen peroxide solution and investigation of its sonocatalytic activity. Ultrason Sonochem 15:301–307

Wang X, Wang J, Guo P, Guo W, Li G (2008e) Chemical effect of swirling jet-induced cavitation: degradation of rhodamine B in aqueous solution. Ultrason Sonochem 15:357–363

Wang X, Yao Z, Wang J, Guo W, Li G (2008f) Degradation of reactive brilliant red in aqueous solution by ultrasonic cavitation. Ultrason Sonochem 15:43–48

Wang Y, Zhao D, Ma W, Chen C, Zhao J (2008g) Enhanced sonocatalytic degradation of azo dyes by Au/TiO$_2$. Environ Sci Technol 42:6173–6178

Wang J, Jiang Z, Zhang Z, Xie Y, Lv Y, Deng Y, Zhang X (2009a) Study on inorganic oxidants assisted sonocatalytic degradation of Acid Red B in presence of nano-sized ZnO powder. Sep Purif Technol 67:38–43

Wang S, Gong Q, Liang J (2009b) Sonophotocatalytic degradation of methyl orange by carbon nanotube/TiO$_2$ in aqueous solutions. Ultrason Sonochem 16:205–208

Wang J, Wang X, Li G, Guo P, Luo Z (2010a) Degradation of EDTA in aqueous solution by using ozonolysis and ozonolysis combined with sonolysis. J Hazard Mater 176:333–338

Wang N, Zhu L, Wang M, Wang D, Tang H (2010b) Sono-enhanced degradation of dye pollutants with the use of H$_2$O$_2$ activated by Fe$_3$O$_4$ magnetic nanoparticles as peroxidase mimetic. Ultrason Sonochem 17:78–83

Wang J, Gui Y, Liu B, Jin X, Liu L, Xu R, Kong Y, Wang B (2011a) Detection and analysis of reactive oxygen species (ROS) generated by nano-sized TiO$_2$ powder under ultrasonic irradiation and application in sonocatalytic degradation of organic dye. Ultrason Sonochem 18:177–183

Wang J, Wang X, Guo P, Yu P (2011b) Degradation of reactive brilliant red K-2BP in aqueous solution using swirling jet-induced cavitation combined with H$_2$O$_2$. Ultrason Sonochem 18:494–500

Wang XK, Wei YC, Wang C, Guo WL, Wang JG, Jiang JX (2011c) Ultrasonic degradation of reactive brilliant red K-2BP in water with CCl$_4$ enhancement: performance optimization and degradation mechanism. Sep Purif Technol 81:69–76

Wang Y, Zhang H, Chen L (2011d) Ultrasound enhanced catalytic ozonation of tetracycline in a rectangular air-lift reactor. Catal Today 175:283–292

Wang Y, Zhang H, Chen L, Wang S, Zhang D (2012) Ozonation combined with ultrasound for the degradation of tetracycline in a rectangular air-lift reactor. Sep Purif Technol 84:138–146

Wu CH (2009) Photodegradation of C.I. Reactive Red 2 in UV/TiO$_2$-based systems: effects of ultrasound irradiation. J Hazard Mater 167:434–439

Wu ZL, Ondruschka B, Cravotto G (2008) Degradation of phenol under combined irradiation of microwaves and ultrasound. Environ Sci Technol 42:8083–8087

Yao JJ, Gao NY, Li C, Li L, Xu B (2010a) Mechanism and kinetics of parathion degradation under ultrasonic irradiation. J Hazard Mater 175:138–145

Yao JJ, Gao NY, Ma Y, Li HJ, Xu B, Li L (2010b) Sonolytic degradation of parathion and the formation of byproducts. Ultrason Sonochem 17:802–809

Yavuz Y, Koparal AS, Artik A, Öğütveren ÜB (2009) Degradation of C.I. Basic Red 29 solution by combined ultrasound and Co^{2+}-H$_2$O$_2$ system. Desalination 249:828–831

Yuan S, Yu L, Wu J, Fang J, Zhao Y (2009) Highly ordered TiO$_2$ nanotube array as recyclable catalyst for the sonophotocatalytic degradation of methylene blue. Catal Commun 10:1188–1191

Zhang H, Lv Y, Liu Y, Zhang D (2008a) Degradation of C.I. Acid Orange 7 by ultrasound enhanced ozonation in a rectangular air-lift reactor. Chem Eng J 138:231–238

Zhang Z, Yuan Y, Liang L, Fang Y, Cheng Y, Ding H, Shi G, Jin L (2008b) Sonophotoelectrocatalytic degradation of azo dye on TiO$_2$ nanotube electrode. Ultrason Sonochem 15:370–375

Zhang H, Fu H, Zhang D (2009a) Degradation of C.I., Acid Orange 7 by ultrasound enhanced heterogeneous Fenton-like process. J Hazard Mater 172:654–660

Zhang H, Zhang J, Zhang C, Liu F, Zhang D (2009b) Degradation of C.I. Acid Orange 7 by the advanced Fenton process in combination with ultrasonic irradiation. Ultrason Sonochem 16:325–330

Zhang K, Gao N, Deng Y, Lin TF, Ma Y, Li L, Sui M (2011a) Degradation of bisphenol-A using ultrasonic irradiation assisted by low-concentration hydrogen peroxide. J Environ Sci 23:31–36

Zhang K, Zhang FJ, Chen ML, Oh WC (2011b) Comparison of catalytic activities for photocatalytic and sonocatalytic degradation of methylene blue in present of anatase TiO_2-CNT catalysts. Ultrason Sonochem 18:765–772

Zhang Y, Hou Y, Chen F, Xiao Z, Zhang J, Hu X (2011c) The degradation of chlorpyrifos and diazinon in aqueous solution by ultrasonic irradiation: effect of parameters and degradation pathway. Chemosphere 82:1109–1115

Zhao J, Wang X, Zhang L, Hou X, Li Y, Tang C (2011) Degradation of methyl orange through synergistic effect of zirconia nanotubes and ultrasonic wave. J Hazard Mater 188:231–234

Zhong X, Royer S, Zhang H, Huang Q, Xiang L, Valange S, Barrault J (2011a) Meosporous silica iron-doped as stable and efficient heterogeneous catalyst for the degradation of C.I. Acid Orange 7 using sono-photo-Fenton process. Sep Purif Technol 80:163–171

Zhong X, Xiang L, Royer S, Valange S, Barrault J, Zhang H (2011b) Degradation of C.I. Acid Orange 7 by heterogeneous Fenton oxidation in combination with ultrasonic irradiation. J Chem Technol Biotechnol 86:970–977

Zhou T, Li Y, Wong FS, Lu X (2008) Enhanced degradation of 2,4-dichlorophenol by ultrasound in a new Fenton like system (Fe/EDTA) at ambient circumstance. Ultrason Sonochem 15:782–790

Zhou T, Lim TT, Lu X, Li Y, Wong FS (2009) Simultaneous degradation of 4CP and EDTA in a heterogeneous Ultrasound/Fenton like system at ambient circumstance. Sep Purif Technol 68:367–374

Zhou T, Lim TT, Li Y, Lu X, Wong FS (2010) The role and fate of EDTA in ultrasound-enhanced zero-valent iron/air system. Chemosphere 78:576–582

Zhou T, Lim TT, Wu X (2011) Sonophotocatalytic degradation of azo dye reactive black 5 in an ultrasounic/UV/ferric system and the roles of different organic ligands. Water Res 45:2915–2924

Zhu C, Chen Z, Ni C, Yu J, Huang B, Shan M (2011) Degradation of sodium polystyrene sulfonate and the radical initiated polymerization of styrene under ultrasonic irradiation. Polym-Plast Technol 50:1262–1265

Zouaghi R, David B, Suptil J, djebbar K, Boutiti A, Guittonneau S (2011) Sonochemical and sonocatalytic degradation of monolinuron in water. Ultrason Sonochem 18:1107–1112

Chapter 4
Efficiency Issues for Ultrasound

Abstract The yield of sonication depends heavily on ultrasonic factors. Hence, the experimental conditions for ultrasound treatment must be carefully considered when a process is designed and controlled during the ultrasonic irradiation. Operating conditions such as applied ultrasound frequency, ultrasound intensity, liquid bulk temperature, initial pH of the solution, initial substrate concentration, and others affect ultrasound treatment performance in a positive or adverse way. As ultrasound alone is usually insufficient for total mineralization of organic compounds in the wastewater, the addition of various additives and combined or integrated treatments are of common interests for improving mineralization reaction and enhancing degradation efficiency of the pollutant as a whole. This chapter is a brief account of the main parameters influencing cavitation chemistry and ways to enhance the ultrasound treatment performance.

Keywords Additives · Initial concentration · Integrated treatments · pH · Temperature · Ultrasound frequency and itensity

4.1 Operating Parameters

4.1.1 Ultrasound Frequency

Generally, the degradation rate is dependent on the number of radicals formed in the bubbles and on the extent of radical released to the bulk liquid (Xie et al. 2011). Higher ultrasound frequencies usually increase the number of free radicals because there are more cavitational events that consequently lead to an increase of pollutant degradation. However, there is an optimum frequency which maximizes the degradation rate of pollutant. When the applied frequency exceeded the

T. Y. Wu et al., *Advances in Ultrasound Technology for Environmental Remediation*,
SpringerBriefs in Green Chemistry for Sustainability, DOI: 10.1007/978-94-007-5533-8_4,
© The Author(s) 2013

optimum frequency, the collapse of the bubbles occurs more rapidly, causing more radicals to escape from the bubbles (Xie et al. 2011). Hartmann et al. (2008) and Isariebel et al. (2009) determined the optimum frequency occurred at 617 and 574 kHz, respectively, giving the highest rate of pollutant degradation. David (2009) observed the highest activity occurred at 506 kHz, as compared to 20 kHz and was tentatively explained by the examination of the physical characteristics of the bubbles as well as the calculation of the number of bubbles at both frequencies. Nevertheless, Sponza and Oztekin (2011) observed that increasing the sonification frequency from 35 to 150 kHz did not increase the degradation ratio of PAHs. Yang et al. (2008) and Deojay et al. (2011) observed that the degradation of pollutant was enhanced under a certain set of pulsing and ultrasound frequency exposure conditions, attributed to the adsorption process with an accumulation of pollutant at the gas/solution interface of cavitation bubbles.

4.1.2 Ultrasound Intensity

The power intensity of ultrasound is defined as the power delivered to the liquid divided by the surface area of the ultrasonic transducer. The relationship between the ultrasonic power intensity and the acoustic pressure is expressed as (Mason and Lorimer 1988)

$$I = \frac{P_o^2}{2\rho C} \tag{4.1}$$

where
I   is the power intensity of a sound wave
P_o is the acoustic pressure
ρ is the density of the liquid
C   is the sound speed in the liquid

This relationship shows that an increase in power intensity of ultrasound will enhance the acoustic pressure, thus encouraging more violent cavitational collapse to be occurred (Chen 2012). Sponza and Oztekin (2011) stated that the ultrasonic irradiation of high output power could facilitate the dispersion of organic pollutants with OH, resulting in the destruction of pollutants. A mixture of PAHs degradation was observed to increase from 60 to 83 % when the power was doubled from 75 to 150 W (Manariotis et al. 2011). Suri et al. (2007), Isariebel et al. (2009), and Naddeo et al. (2010) also reported similar trends where higher degradation rates or efficiencies was obtained when the power intensity was increased.

Although an increase of power dissipation in the ultrasound system could improve the degradation of pollutants, the rates of degradation may decrease with further increase in power which is beyond its optimum value (Isariebel et al. 2009,

Guo et al. 2010b). For example, ultrasound power of 400 W was observed to be the optimum power for achieving maximum removal rate of Levofloxacin (Guo et al. 2010b). This is attributed to the large number of gas bubbles existed in the solution at higher input power (exceeded the optimum power) which could scatter the sound waves to the wall of the vessel, thus lowering the energy dissipated in the liquid (Guo et al. 2010a, b). Chen and Huang (2009) monitored the degradation rate of nitrotoluenes. They found that the degradation rate increased proportionately with an increase in power intensity up to 102 W/cm^2 but higher power intensity reduced the degradation rate.

4.1.3 Temperature

In general, an elevation of reaction temperature can enhance pollutant removal efficiency as higher temperature facilitates the bubble formation and allows the pollutant molecules to move faster into the cavitation bubbles. However, there is an optimum reaction temperature in which the maximum degradation efficiency is achieved and the degradation starts decreasing once reaction temperature exceeded optimum temperature (Sponza and Oztekin 2011). However, Chen and Huang (2009, 2011) found that TOC removal rate was higher at lower temperature, which could be attributed to suppression in cavitation intensity due to an increase of solvent vapor pressure caused by the elevation of reaction temperature. Li et al. (2008) also obtained similar trend and concluded that elevated temperature facilitated the loss of cavitation energy and the cavitation bubble generated by ultrasonic wave escaped more easily at higher temperature.

4.1.4 Initial pH

Initial pH of the medium plays an important role for the degradation of chemical pollutants under ultrasound irradiation. The effect of pH on the degradation rate depends strongly on the state of the pollutant molecule (i.e., the pollutant present as ionic species or as a molecule). With an extreme pH condition (very high or very low pH), the production of OH and H increases, thereby making chemical effects dominant in the degradation of pollutant (Vijayalakshmi and Madras 2006). For example, the degradation of ciprofloxacin at pH 3 was observed to be four times faster than at pH 7 (Bel et al. 2011). This phenomenon was explained by the degree of protonation of ciprofloxacin and positive charges on the ciprofloxacin molecule seemed to promote ultrasonic degradation, due to the accumulation at the negatively charged liquid–bubble interface (Bel et al. 2011). Decoloration efficiency increased at acidic condition, which was probably associated with the effect of protonation of negative charges—SO_3^- groups in acidic medium and the hydrophobic character of the resulting molecule enhanced its reactivity under sonochemical process (Vajnhandl and Marechal 2007). Similar trend was observed by Shriwas and Gogate (2011) where maximum degradation of pollutant was achieved at pH 2.5, and a significant decrease in degradation was obtained when the pH increased to pH 6. Gűltekin and Ince (2008) reported that the decomposition of bisphenol-A at pH 3 and 6 were faster than at pH 10.5, which could be explained as the easier diffusion of the molecule at lower pH to the bubble–liquid interface, where the concentration of OH· was a maximum. However, Jiang et al. (2002) observed the degradation rate of aniline under ultrasonic irradiation was lower at pH 4 than in alkaline solution (pH 8.11), which was attributed to the high solubility of the ionic anilinium ion and the preferential movement of the uncharged form to the interface.

4.1.5 Initial Concentration

Commonly, an increase of initial concentration of the pollutants reduces the degradation efficiency of ultrasonic irradiation. For example, Kritikos et al. (2007), Vajnhandl and Marechal (2007), and He et al. (2008) indicated that the degree of decoloration or mineralization of dye increased with decreasing initial concentration. This is attributed to two possible reasons: (1) the cavities and OH approached saturation gradually with an increase in initial dye concentration, and (2) higher initial dye concentration resulted in generation of more inorganic anions, which competed with carbonaceous organic substances for reaction with OH· (He et al. 2008). Similar trend was observed by Güyer and Inch (2011). Nevertheless, Madhavan et al. (2010) observed the removal rate of diclofenac increased with an increase of diclofenac concentration until 0.05 mM and then it leveled off, which was explained by considering the formation of OH· through sonolysis of water.

4.2 Integrated Treatments with Other Advanced Oxidation Processes

Studies on the ultrasound irradiation of pollutants have demonstrated its potential for decomposition and degradation of pollutants in wastewater. Nevertheless, the ultrasonic irradiation alone is usually not enough to provide high degradation rate to be used practically. Therefore, integration or combined ultrasound application with other Advanced Oxidation Processes (AOPs) is one of the solutions to increase the degradation efficiency (Liang et al. 2007).

4.2.1 Ultrasound/O_3

When ozone is injected into water along with ultrasound, an additional pathway of OH generation arises upon the decomposition of ozone in the gaseous bubbles during implosive collapse. Therefore, this combination could enhance the generation of additional free radicals. The mechanism of the effect of ultrasound/O_3 treatment can be represented by Eqs. 4.1–4.3 (Kidak and Ince 2007).

$$H_2O \; + \; \text{Ultrasound} \rightarrow OH\cdot + H \tag{4.2}$$

$$O_3(g) \; + \; \text{Ultrasound} \rightarrow O_2(g) + \; O(^3P)\,(g) \tag{4.3}$$

$$O(^3P)\,(g) + \; H_2O\;(g) \; + \; \text{Ultrasound} \rightarrow 2OH\cdot\,(g) \tag{4.4}$$

The degradation of p-aminophenol was investigated using ultrasound-enhanced ozonation (He et al. 2007b). The degradation efficiency of pollutant using combination treatments (99 %) exceeded the efficiency of using ultrasound alone (4 %). In addition, TOC removal after 720 min of reaction was higher by using the combined process (77 %) as compared to a single ultrasound process (8 %) (He et al. 2007b). About 21.9 % degradation of pollutant was obtained in presence of O_3 as compared to only 4.1 % without any bubbling ozone through solution (Shriwas and Gogate 2011). Bisphenol was completely removed after 60 min treatment in the combined ultrasound/O_3 system, whereas only 34.6 % was achieved with a single ultrasound treatment (Guo and Feng 2009). Kidak and Ince (2007) obtained similar results as rate of degradation was larger with ultrasound/O_3 treatment than in a single ultrasound operation. This combined system also resulted in complete degradation of oxalate and TOC under an hour treatment. The oxidation rates could be enhanced 16 times greater than a simple linear addition of the two independent reaction systems (Vecitis et al. 2010). These results were concurrent with the studies conducted by He et al. (2007a, b).

4.2.2 Ultrasound/UV (Sonophotolytic)

Integration of ultrasound with UV was found to provide a considerable advantage over single ultrasound operation by the formation of excess OH· upon photolysis of ultrasound-induced H_2O_2 (Kidak and Ince 2007). Ultrasound/UV combination doubled the phenol degradation rate as compared to using ultrasound alone but this rate decelerated by pH elevations (Kidak and Ince 2007). Aarthi et al. (2007) explained that an increase in degradation using the combined ultrasound/UV process was because an increase in the number of scission products per breakage and not due to the increase in the intrinsic rate. Rashid and Sato (2011) showed that the process using ultrasound/UV exhibited larger degradation efficiencies than using ultrasound alone.

4.2.3 Ultrasound/UV/Photocatalysts (Sonophotocatalytic)

When two modes of irradiations (UV and ultrasound) are operated in combinations, more number of free radicals will be available for the reaction, thereby increasing the rates of reaction (Gogate 2008). Mishra and Gogate (2011) compared the efficiency of combined process, namely sonophotocatalytic (TiO$_2$ as photocatalyst) with an individual operation of sonolysis. Their study showed that the pollutant was effectively removed by using sonophotocatalytic treatment (65.4 % degradation) and the degradation rate could be enhanced by the adding additives such as H$_2$O$_2$ (94.6 % degradation, with 1 g/L H$_2$O$_2$) due to the generation of enhanced quantum of OH· (Mishra and Gogate 2011). Shriwas and Gogate (2011) reported similar observations where 20.9 % of the pollutant removal was achieved using the combination process as compared to 4.1 % by using ultrasound sonification alone. Similar trends were obtained by Kaur and Singh (2007), Kritikos et al. (2007), Gonzalez and Martinez (2008), Wu (2009), and others. Bi$_2$O$_3$/TiZrO$_4$ was applied as a visible light-driven photocatalyst together with ultrasound irradiation to study the degradation of 4-chlorophenol. This study indicated that higher degradation efficiency of pollutant was obtained during the sonophotocatalytic process (75 %) as compared to sonolysis alone (63 %) (Neppolian et al. 2011).

4.2.4 Ultrasound/ Fe/EDTA (Fenton-Like System)

The combination process of ultrasound/Fe/EDTA saves treatment cost remarkably by using iron metal instead of ferrous salt (Zhou et al. 2008). EDTA was found to break down O–O bond of oxygen and eventually produce H$_2$O$_2$ in a Fe/EDTA system, and was very attractive due to the adoption of low cost "green" EDTA with oxygen to generate H$_2$O$_2$. Complete degradation of pollutant was achieved using this combination system as compared to 33.5 % removal of pollutant in ultrasound alone (Zhou et al. 2008). Zhou et al. (2009) obtained similar results, in which rapid and high removal percentage (100 %) of pollutant was achieved in the ultrasound/Fe/EDTA system. Theoretical mechanism of this combination process is summarized as shown:

$$\mathrm{Fe^{2+} + EDTA \longrightarrow} k_1 \left[\mathrm{Fe^{II}(EDTA)} \right] \tag{4.5}$$

$$\left[\mathrm{Fe^{II}(EDTA)(H_2O)} \right]^{2-} + \mathrm{O_2} \overset{k_2, k_{-2}}{\longleftrightarrow} \left[\mathrm{Fe^{II}(EDTA)(O_2)} \right]^{2-} + \mathrm{H_2O} \tag{4.6}$$

$$\left[\mathrm{Fe^{II}(EDTA)(O_2)} \right]^{2-} \longrightarrow k_3 \left[\mathrm{Fe^{III}(EDTA)(O_2^-)} \right]^{2-} \tag{4.7}$$

$$\left[\mathrm{(EDTA)Fe^{III}(O_2^{2-})Fe^{III}(EDTA)} \right]^{4-} \longrightarrow k_5, H^+ 2\left[\mathrm{Fe^{III}(EDTA)H_2O} \right]^- + \mathrm{H_2O_2} \tag{4.8}$$

$$\left[(EDTA)Fe^{III}(O_2^{2-})Fe^{III}(EDTA)\right]^{4-} \longrightarrow k_5, H^+ 2\left[Fe^{III}(EDTA)H_2O\right]^- + H_2O_2$$
(4.9)

4.2.5 Ultrasound/H_2O_2/Fe_2O_3 (Sono-Fenton)

In Sono-Fenton reaction, iron metal is initially corroded in the presence of H_2O_2 under acidic conditions, oxidizing Fe^0 to Fe^{2+}, which then further reacts with H_2O_2 in the process to generate OH· and Fe^{3+}. The reactions that occur in the system are given as shown:

$$Fe^0 + 2H^+ \rightarrow Fe^{2+} + H_2$$
(4.10)

$$Fe^{2+} + H_2O_2 \rightarrow Fe^{3+} + OH^- + OH·$$
(4.11)

$$H_2O_2 + Ultrasound \rightarrow 2OH·$$
(4.12)

$$H_2O_2 + OH· \rightarrow H_2O + OOH$$
(4.13)

$$Fe^{3+} + H_2O_2 \rightarrow Fe(OOH)^{2+} + H^+$$
(4.14)

$$Fe(OOH)^{2+} + Ultrasound \rightarrow Fe^{2+} + OOH$$
(4.15)

Fe-containing catalyst such as iron powder and mill scale was observed to enhance the degradation of 4-chlorophenol when it was reacted with H_2O_2 under ultrasound irradiation whereas basic oxygen furnace slag showed no catalysis effect of the degradation (Liang et al. 2007). Complete degradation of the pollutant was observed within 2 min of ultrasonic irradiation with an addition of 1 g/L of iron powder or mill scale (Liang et al. 2007). A combination of Sono-Fenton process and cavitation has been observed to intensity the degradation process by way of turbulence and generation of additional free radicals. Bremner et al. (2008) reported that when an aqueous solution of 2,4-dichlorophenoxyacetic was rapidly stirred with zero-valent iron powder in the presence of H_2O_2 under ultrasound process, a rapid degradation of pollutant was observed but leveled off after 20 min. More than 99 % degradation and 46 % mineralization of carbofuran were achieved after undergoing 30 min reaction time at pH 3 for an initial concentration of 20 mg/L, together with 100 mg/L H_2O_2 and 20 mg/L Fe^{2+} (Ma et al. 2010). These results were in agreement with Sun et al. (2007), Hsieh et al (2009) as well as Li and Song (2010).

4.3 Additives

Although organic compounds can be decomposed through sonolysis, the degradation rates are still low for real practical use. Recently, many studies have demonstrated that an addition of additives such as carbon tetrachloride, sodium chloride, surfactant, and others help to increase the extent of degradation and mineralization.

4.3.1 Carbon Tetrachloride (CCl₄)

Several studies indicated that addition of CCl_4 enhanced the ultrasound degradation rate and reduced the time required for removing the pollutants. This is attributed to its high volatility property, which facilitates its diffusion into gaseous bubble interior to undergo molecular fragmentation, releasing oxidizing agents that could react with organic molecules (Guo et al. 2010b). Laxmi et al. (2010) suggested that CCl_4 acted as a H scavenger and a hydrophobic organic compound, which was prone to enter the cavitation bubbles and to be degraded by pyrolytic cleavage. (Guo et al. 2010b) reported that the removal percentage of levoflaxin increased from 9.4 to 94.8 % after 20 min of ultrasonic irradiation in the presence of 0.02 mL CCl_4. Similar trends were observed by Guo et al. (2008, 2010a) in which the removal of 2,4-dinitrophenol and levofloxacin increased significantly after an addition of CCl_4. By adding CCl_4 into the ultrasonic system, the percentage degradation of 2chloro-5methy phenol was increased significantly from 12 % (ultrasonic irradiation) to 33 % (ultrasound/CCl_4) (Laxmi et al. 2010).

4.3.2 Catalysts (TiO₂, SiO₂, SnO₂)

Hartmannet et al. (2008) confirmed that the presence of catalysts accelerated the degradation of diclofenac because the relative concentration of diclofenac was reduced to 16% of the initial diclofenac concentration with an addition of TiO_2. The increase in degradation rate was explained by the fragmentation of catalyst through cavitation process which produced higher surface area (Laxmi et al. 2010). Therefore, ultrasonic irradiation not only destroys pollutant, but also increases the adsorption process by increasing the surface area of the catalyst (Laxmi et al. 2010). Shriwas and Gogate (2011) obtained similar results with an addition of TiO_2 as 20 % of TOC removal was achieved. Shimizu et al. (2008) evaluated the effect of TiO_2 on an ultrasonic system's oxidation power is evaluated by examining the oxidation of salicylic acid. Their study concluded that the presence of TiO_2 accelerated the generation of OH· during sonolysis. It was shown by Zouaghi et al. (2011) that ultrasound efficiency was improved in the presence of nanoparticles of TiO_2 and SiO_2 at 20 kHz because they provided nucleation sites for cavitation bubbles at their surface.

4.3.3 Fe-Containing Additives (ZVI, DVI, NPI)

An application of ultrasound along with Fe-containing additives can enhance the mass transport of reactants to the metal surface, where ultrasound increases the defects and the number of active sites, while continuously cleaning it (Güyer and Ince 2011). However, Güyer and Ince (2011) found that for each additive, there existed a threshold concentration, above which the efficiency of degradation would be reduced. Application of zero valent iron (ZVI) was investigated by Yang et al. (2010) to accelerate the degradation of α-,β-, γ-, and δ-hexachlorocyclohexane (HCHs) and DDX (DDT, DDE and DDD) in the soil from a former organochlorine pesticide manufacturing plant. The results showed that zero valent iron could facilitate the degradation of β-HCH, p,p'-DDT, and o,p'-DDT, but had little effect on the degradation of α-HCH, γ-HCH, and δ-HCH (Yang et al. 2010).

4.3.4 Salt

Sivasankar and Moholkar (2009) observed that the degradation of pollutant increased with an addition of salt. This phenomenon was explained by the hydrophobic repulsive interaction between pollutant and water molecules, where the pollutant molecules were "pushed" toward the bubble interface, causing higher concentration of pollutant molecules. Thus, it enhanced the probability of radical–pollutant interaction as well as the extent of pollutant evaporation into the bubble due to rise in the partial pressure of the pollutant at the bubble interface. However, this effect was marked only for the pollutants with hydrophilic character. If the pollutant exhibited a strong hydrophobic character, its concentration at the bubble-bulk interfacial region was already at saturation and would not change with salt addition (Sivasankar and Moholkar 2009). The extent of phenol degradation was also found to be enhanced from 9.5 to 14.5 % with the addition of 1.5 g/L NaCl (Khokhawala and Gogate 2011). Similar result was obtained when two types of salts (NaCl and $NaNO_2$) were applied by Katekhaye and Gogate (2011) to intensify the cavitational activity.

4.3.5 Coal Ash

Nakui et al. (2007) indicated that 0.4–0.6 wt% coal ash accelerated the phenol degradation due to an increase in the amount of OH under the ultrasonic irradiation. Since the coal ash used had a porous and uneven surface, it was assumed that the coal ash led to the increase in the nucleation site for cavitation bubble due to its surface roughness. However, at larger amount of coal ash (>0.6 wt%), ultrasonic wave would begin to be scattered or absorbed by the coal ash, by which the formation of the nucleation site for the cavitation bubble was assumed to become difficult (Nakui et al. 2007). Similar effect was observed by Liu et al. (2009) because coal ash acted as a catalyst to generate OH with the presence of H_2O_2 or O_3. In the presence of coal ash, up to 83.4 and 88.8 % degradation of phenol were achieved with H_2O_2 and O_3, respectively, under sonochemical treatment.

4.4 Conclusion

Although organic compounds can be decomposed through sonolysis, the degradation rates are still low for real practical use. Besides, ultrasound is an AOP process that concentrates the hydrophobic substrate and OH at the bubble–solution interface. However, due to the formation of highly hydrophilic by-products during the process, poor mineralization is usually observed using this technology. Recent research focuses on finding new mechanisms and reaction pathways in order to enhance the ultrasound energy efficiency, with the combination of sonolysis with other AOPs (such as ozonolysis, Fenton's reaction, and others) and/or with an addition of additive such as CCl_4, catalysts, Fe-containing additives, salt, and coal ash.

References

Aarthi T, Shaama MS, Madras G (2007) Degradation of water soluble polymers under combined ultrasonic and ultraviolet radiation. Ind Eng Chem Res 46:6204–6210
Bel ED, Janssen C, Smet SD, Langenhove HV, Dewulf J (2011) Sonolysis of ciprofloxacin in aqueous solution: Influence of operational parameters. Ultrason Sonochem 18:184–189
Bremner DH, Carlo SD, Chakinala AG, Cravotto G (2008) Mineralisation of 2,4-dichlorophenoxyacetic acid by acoustic or hydrodynamic cavitation in conjunction with the advanced Fenton process. Ultrason Sonochem 15:416–419
Chen D (2012) Applications of ultrasound in water and wastewater treatment. In: Chen D, Sharma SK, Mudhoo A (eds) Handbook on application of ultrasound: sonochemistry for sustainability. CRC Press, Taylor & Francis Group, Boca Raton
Chen WS, Huang GC (2009) Sonochemical decomposition of dinitrotoluenes and trinitrotulene in wastewater. J Hazard Mater 169:868–874
Chen WS, Huang YL (2011) Removal of dinitrotoluenes and trinitrotoluene from industrial wastewater by ultrasound enhanced with titanium dioxide. Ultrason Sonochem 18:1232–1240
David B (2009) Sonochemical degradation of PAH in aqueous solution. Part I: monocomponent PAH solution. Ultrason Sonochem 16:260–265

Deojay DM, Sostaric JZ, Weavers LK (2011) Exploring the effects of pulsed ultrasound at 205 and 616 kHz on the sonochemical degradation of octylbenzene sulfonate. Ultrason Sonochem 18:801–809

Gogate PR (2008) Treatment of wastewater streams containing phenolic compounds using hybrid techniques based on cavitation: a review of the current status and the way forward. Ultrason Sonochem 15:1–15

Gonzalez AS, Martinez SS (2008) Study of the sonophotocatalytic degradation of basic blue 9 industrial textile dye over slurry titanium dioxide and influencing factors. Ultrason Sonochem 15:1039–1042

Gültekin I, Ince NH (2008) Ultrasonic destruction of bisphenol-A: the operating parameters. Ultrason Sonochem 15:524–529

Guo Z, Feng R (2009) Ultrasonic irradiation-induced degradation of low-concentration bisphenol A in aqueous solution. J Hazard Mater 163:855–860

Guo Z, Feng R, Li J, Zheng Z, Zheng Y (2008) Degradation of 2,4-dinitrophenol by combining sonolysis and different additives. J Hazard Mater 158:164–169

Guo W, Shi Y, Wang H, Yang H, Zhang G (2010a) Intensification of sonochemical degradation of antibiotics levofloxacin using carbon tetrachloride. Ultrasound Sonochem 17:680–684

Guo W, Shi Y, Wang H, Yang H, Zhang G (2010b) Sonochemical decomposition of levofloxacin in aqueous solution. Water Environ Res 82:696–700

Güyer GT, Ince NH (2011) Degradation of diclofenac in water by homogeneous and heterogeneous sonolysis. Ultrason Sonochem 18:114–119

Hartmann J, Bartels P, Mau U, Witter M, Tumpling WV, Hofmann J, Nietzschmann (2008) Degradation of the drug diclofenac in water by sonolysis in presence of catalyst. Chemosphere 70:453–461

He Z, Song S, Xia M, Qiu J, Ying H, Lu B, Jiang Y, Chen J (2007a) Mineralization of C.I. reactive yellow 84 in aqueous solution by sonolytic ozonation. Chemosphere 69:191–199

He Z, Song S, Ying H, Xu L, Chen J (2007b) p-Aminophenol degradation by ozonation combined with sonolysis: operating conditions influence and mechanism. Ultrason Sonochem 14:568–574

He Z, Lin L, Song S, Xia M, Xu L, Ying H, Chen J (2008) Mineralization of C.I. Reactive Blue 19 by ozonation combined with sonolysis: performance optimization and degradation mechanism. Sep Purif Technol 62:376–381

Hsieh LL, Kang HJ, Shyu HL, Chang CY (2009) Optimal degradation of dye wastewater by ultrasound/Fenton method in the presence of nanoscale iron. Water Sci Technol 60:1295–1301

Isariebel QP, Carine JL, Ulises-Javier JH, Anne-Marie W, Henri D (2009) Sonolysis of levodopa and paracetamol in aqueous solutions. Ultrason Sonochem 16:610–616

Jiang Y, Pétrier C, Waite TD (2002) Effect of pH on the ultrasonic degradation of ionic aromatic compounds in aqueous solution. Ultrason Sonochem 9:163–168

Katelhaye SN, Gogate PR (2011) Intensification of cavitational activity in sonochemical reactors using different additives: efficacy assessment using a model reaction. Chem Eng Process 50:95–103

Kaur S, Singh V (2007) Visible light induced sonophotocatalytic degradation of Reactive Red dye 198 using dye sensitized TiO_2. Ultrason Sonochem 14:531–537

Kidak R, Ince NH (2007) Catalysis of advanced oxidation reactions by ultrasound: a case study with phenol. J Hazard Mater 146:630–635

Kokhawala IM, Gogate PR (2011) Intensification of sonochemical degradation of phenol using additive at pilot scale operation. Water Sci Technol 63:2547–2552

Kritikos DE, Xekoukoulatakis NP, Psillakis E, Mantzavinos (2007) Photocatalysitc degradation of reactive black 5 in aqueous solutions: effect of operating conditions and coupling with ultrasound irradiation. Water Res 41:2236–2246

Laxmi PNV, Saritha P, Rambabu N, Himabindu V, Anjaneyulu Y (2010) Sonochemical degradation of 2chloro-5 methyl phenol assisted by TiO_2 and H_2O_2. J Hazard Mater 174:151–155

Li JT, Song YL (2010) Degradation of AR 97 aqueous solution by combination of ultrasound and Fenton reagent. Environ Prog Sustain 29:101–106

Li J, Cai J, Fan L (2008) Effect of sonolysis on kinetics and physiochemical properties of treated chitosan. J Appl Polym Sci 109:2417–2425

Liang J, Komarov S, Hayashi N, Kasai (2007) Improvement in sonochemical degradation of 4-chlorophenol by combined use of Fenton-like reagents. Ultrason Sonochem 14:201–207

Liu H, Liang MY, Liu CS, Gao YX, Zhou JM (2009) Catalytic degradation of phenol in sonolysis by coal ash and H_2O_2/O_3. Chem Eng J 153:131–137

Ma YS, Sung CF, Lin JG (2010) Degradation of carbofuran in aqueous solution by ultrasound and Fenton processes: effect of system parameters and kinetic study. J Hazard Mater 178:320–325

Madhavan J, Kumar PSS, Anandan S, Zhou M, Grieser F, Ashokkumar M (2010) Ultrasound assisted photocatalytic degradation of diclofenac in an aqueous environment. Chemosphere 80:747–752

Manariotis ID, Karapanagioti HK, Chrysikopoulos CV (2011) Degradation of PAHs by high frequency ultrasound. Water Res 45:2587–2594

Mason TJ, Lorimer JP (1988) Sonochemistry: theory, applications and uses of ultrasound in chemistry. Ellis Horwood, Chichester

Mishra KP, Gogate PR (2011) Intensification of sonophotocatalytic degradation of p-nitrophenol at pilot scale capacity. Ultrason Sonochem 18:739–744

Naddeo V, Belgiorno V, Kassinos D, Mantzavinos D, Meric S (2010) Ultrasonic degradation, mineralization and detoxification of diclofenac in water: optimization of operating parameters. Ultrason Sonochem 17:179–185

Nakui H, Okitsu K, Maeda Y, Nishimura R (2007) Effect of coal ash on sonochemical degradation of phenol in water. Ultrason Sonochem 14:191–196

Neppolian B, Ciceri L, L. Bianchi C, Grieser F, Ashokkumar M (2011) Sonophotocatalytic degradation of 4-chlorophenol using $Bi_2O_3/TiZrO_4$ as a visible light responsive photocatalyst. Ultrason Sonochem 18:135–191

Rashid MM, Sato C (2011) Photolysis, sonolysis and photosonolysis of trichloroethane (TCA), trichloroethylene (TCE), and tetrachloroethylene (PCE) without catalyst. Water Air Soil Pollut 216:429–440

Shimizu N, Ogino C, Dadjour MF, Ninomiya K, Fujihira A, Sakiyama K (2008) Sonocatalytic facilitation of hydroxyl radical generation in the presence of TiO_2. Ultrason Sonochem 15:988–994

Shriwas AK, Gogate PR (2011) Intensification of degradation of 2,4,6-trichlorophenol using sonochemical reactors: understanding mechanism and scale-up aspects. Ind Eng Chem Res 50:9601–9608

Sivasankar T, Moholkar VS (2009) Physical insights into the sonochemical degradation of recalcitrant organic pollutants with cavitation bubble dynamics. Ultrason Sonochem 16:769–781

Sponza DT, Oztekin R (2011) Effect of ultrasonic irradiation on the treatment of poly-aromatic substances (PAHs) from a petrochemical industry wastewater. Ozone-Sci Eng 33:192–210

Sun JH, Sun SP, Sun JY, Sun RX, Qiao LP, Guo HQ, Fan MH (2007) Degradation of azo dye acid black 1 using low concentration iron of Fenton process facilitated by ultrasonic irradiation. Ultrason Sonochem 14:761–766

Suri RPS, Nayak M, Devaiah U, Helmig E (2007) Ultrasound assisted destruction of estrogen hormones in aqueous solution: Effect of power density, power intensity and reactor configuration. J Hazard Mater 146:472–478

Vajnhandl S, Marechal AML (2007) Case study of the sonochemical decolouration of textile azo dye Reactive Black 5. J Hazard Mater 141:329–335

Vecitis CD, Lesko T, Colussi AJ, Hoffmann MR (2010) Sonolytic decomposition of aqueous bioxalate in the presence of ozone. J Phys Chem A 114:4968–4980

Vijayalakshmi SP, Madras G (2006) Effects of the pH, concentration, and solvents on the ultrasonic degradation of poly(vinyl alcohol). J Appl Polym Sci 100:4888–4892

Wu CH (2009) Photodegradation of C.I. Reactive Red 2 in UV/TiO$_2$-based systems: effects of ultrasound irradiation. J Hazard Mater 167:434–439

Xie W, Qin Y, Liang D, Song D, he D (2011) Degradation of m-xylene solution using ultrasonic irradiation. Ultrason Sonochem 18:1077–1081

Yang L, Sostaric JZ, Rathman JF, Weavers LK (2008) Effect of ultrasound frequency on pulsed sonolytic degradation of octylbenzene sulfonic acid. J Phys Chem B 112:852–858

Yang S-C, Lei M, Chen T-B, Li XY, Liang Q, Ma C (2010) Application of zerovalent iron (Fe^0) to enhance degradation of HCHs and DDX in soil from a former organochlorine pesticides manufacturing plant. Chemosphere 79:727–732

Zhou T, Li Y, Wong FS, Lu X (2008) Enhanced degradation of 2,4-dichlorophenol by ultrasound in a new Fenton like system (Fe/EDTA) at ambient circumstance. Ultrason Sonochem 15:782–790

Zhou T, Lim TT, Lu X, Li Y, W FS (2009) Simultaneous degradation of 4CP and EDTA in a heterogeneous Ultrasound/Fenton like system at ambient circumstance. Sep Purif Technol 68:367–374

Zouaghi R, David B, Suptil J, djebbar K, Boutiti A, Guittonneau S (2011) Sonochemical and sonocatalytic degradation of monolinuron in water. Ultrason Sonochem 18:1107–1112

Chapter 5
Challenges and Recent Developments of Sonochemical Processes

Abstract Despite ultrasound technique being one of the "green" technologies in environmental remediation and with many possible diverse field applications, there are hardly any physicochemical transformations carried out in industrial scale of operation due to the lack of unified design and scale-up strategies. Issues in scaling up of sonoreactors to meet industrial needs such as process efficiency and rates, energy conversion, high volume processes, and others present a considerable challenge toward further development of this technique. It is important to ensure that maximum efficiency can be attained in the design of industrial-scale sonoreactors due to the difficulty in replicating the exact reactor geometry and sonochemistry environment similar to laboratory-scale reactors as acoustic cavitation near ultrasonic transducers are relatively higher. Some design improvements to be investigated include transducer arrays and a larger exposed surface for ultrasound source, continuous flow reactor designs, and stirring during sonication. This chapter aims to identify some of the key issues in sonochemical processes for industrial-scale application and to update on some of the recent designs in sonochemical reactors.

Keywords Challenges · Design · Developments · Scale-up · Sonochemical reactor · Transducer

5.1 Challenges in Sonochemical Processes for Pollutant Degradation

In the field of sonochemistry for industrial applications, energy efficiency and scaling up process are the two major challenges faced by the researches (Bizzi et al. 2011). With many studies done, total mineralization of organic pollutants using ultrasound irradiation alone still remains a difficult task as degradation rates

T. Y. Wu et al., *Advances in Ultrasound Technology for Environmental Remediation*, SpringerBriefs in Green Chemistry for Sustainability, DOI: 10.1007/978-94-007-5533-8_5, © The Author(s) 2013

are rather slow for practical application, especially for non-volatile compounds (Grčić et al. 2010, Wang et al. 2011). When the pollutant concentration is very high, barrier to nucleation created by the increase of fluid viscosity must be overcome. The presence of viscous bubbles may also hinder the transport of water vapor, dissolved gases, and volatile organics into bubbles during rectified diffusion processes. This phenomenon will then reduce OH· production and pollutant degradation rate. Therefore, a big challenge lies in providing energy to reach nucleation threshold especially at high frequencies (Thangavadivel et al. 2011).

In addition to slow rate or oxidative potential, ultrasonic irradiation also has the tendency to form harmful by-products if it is used alone. If ultrasound is used, a quick mineralization of the organic compound should be the goal to minimize the existence time of toxic intermediates (Adewuyi 2005). Hence, efforts has been focusing on investigating the combination of various technology with ultrasonic systems in order to achieve a desired efficiency of substrate, total mineralization, and reduction of reaction time required to remove pollutants (Grčić et al. 2010). The design of such hybrid systems often depends on choosing processes that complement each other and leading to a synergistic effect. With economic, physical, and technological limitations of ultrasound system, combinations with other treatment processes could effectively treat resistant wastes and reduce treatment costs substantially over single-step process (Adewuyi 2005). The addition of various additives is of common interest for improving mineralization reactions (Madhavan et al. 2010; Naddeo et al. 2010).

With many promising application of sonochemical reactors, there have been minimal successful applications in industrial-scale operations (Katekhaya and Gogate, 2011). Ultrasonic transducers available are made of limited materials and its selection has to be compatible with the pH of operating conditions. Because of pitting issues, the transducer tips or cone will need to be changed frequently (Thangavadivel et al. 2011). Besides addressing the limitation on physical equipments, there is a need of experiments conducted on different scales of operation in order to fully understand and address issues related to scaling up such as alteration in the flow field and turbulence characteristics (Gogate 2007). Gogate and Pandit (2004) presented some useful design equations which served as an important starting point to establish design/scale-up strategies. However, there are still challenges in converting the design equations into generalized forms which are valid for different reactions and sonochemical reactor configurations. For the development of sonoreactors, they suggested several future works which should be focussed:

(a) Validating similar design equations over a wider range of operating parameters.
(b) Exact quantification of the number of free radicals generated during the collapse and more importantly, the actual number of radicals taking part in the reaction.
(c) Estimating the size of the nuclei/cavity.
(d) Measuring the total bubble activity/transient bubble activity in different grades of violence.

(e) Combining the effect of collapse pressure generated and the maximum bubble size reached.

Furthermore, ultrasonic treatment also faces challenges to meet industrial needs in terms of volumetric flow rate, reaction energy rates, and overall cost (Adewuyi 2005). Mahamuni and Adewuyi (2010) reported that the cost of ultrasound treatment of various pollutants is higher as compared to other treatments using advanced oxidation processes (AOPs). Chen and Huang (2011) supported this statement as they found out from their study that ultrasound treatment alone was not a comparatively efficient method in consideration of energy consumption. It was reported that the overall energy transfer efficiency of this treatment was below 10 % as electrical energy must be converted to mechanical vibration, then to cavitation energy before the degradation of pollutant could take place (Thangavadivel et al. 2011).

5.2 Recent Developments in Design and Scale-Up of Sonoreactors

Sonochemical treatment has the potential to be used as one of the methods for environmental remediation as demonstrated in several studies (Adewuyi 2005). The effects caused by acoustic cavitation phenomena has showed success in a number of applications on laboratory scale but a well-defined design and scale-up methodology are still lacking (Gogate et al. 2011). Existing information available based on the laboratory scale may give a very large scale-up ratios and hence, a very high degree of uncertainty (Gogate 2007). The success of scaling up laboratory ultrasonic process was also impeded by the lack of expertise required in diverse fields such as material science, acoustics, chemical engineering, and others (Sutkar and Gogate 2009).

It is important to recognize that the main issue for scaling up the sonochemical processes is the efficiency. When ultrasound is used alone as a sole treatment process, it is a highly energy-intensive process since not all the cavitational energy produces chemical or physical effects (Adewuyi 2005). Cavitational activity in sonochemical reactor does not only depend on reactor's configurations such as location of transducers, surface area of irradiative element, and dimension of the reactor, but also the operating parameters such as power density, height of liquid medium, and others. Bulk temperature, acoustic intensity, and static pressure in the fluid will also affect the reaction mechanism and the overall yield of sonochemical reaction (Sutkar and Gogate 2009). In addition to increase the efficiency of energy conversion, extensive work is also required to increase the cavitation sites and the rates of free radical generation for a given energy input. Generally, utilization of other energy or an addition of free radicals such as H_2O_2, O_3, air, and ferrous ions is needed to improve the degradation process through ultrasonic irradiation (Adewuyi 2005).

Two most common ultrasonic configuration, ultrasonic horn or ultrasonic bath with a single transducer could not be readily used at large-scale processing due to its very low cavitationally active volume (Bhirud et al. 2004). Maximum cavitation events only occur near to the irradiating surface and there will be a wide variation of energy dissipation rates in the bulk of the liquid (Sutkar and Gogate 2009). Hence, there were designs based on the use of multiple transducers operating at the same or different frequencies or the use of a single large transducer located at the bottom of the reactor which emits sound wave longitudinally away from the bottom perpendicular to the axis of the horn (Bhirud et al. 2004; Kumar et al. 2007). However, the use of multiple transducers did not prove to be practical and economically effective in large industrial installations (Gallego-Juárez et al. 2010).

Gogate et al. (2011) basically summarized some of the important considerations for reactor choice, scale up, and optimization. For reactor choices, Mason (1992) was among the earliest to assess the potential and practical use of ultrasound technology in industries. It was concluded that upon considering a sonoreactor, it is important to identify the type of ultrasonic treatment required for a particular chemical process (Mason 1992). In general, there are two types of ultrasonic applications: those based on chemical effect (sonochemistry) and those based on physical effects from bubble collapse (sonoprocessing) (Toma et al. 2011). For sonochemistry, cavitation must be provided during the transformation itself either in a continuous manner or in suitable pulsed operation (Gogate et al. 2011). This effect is caused by the production of OH· and H·, which in turn generate or influence some chemical reaction. On the other hand, sonoprocessing is related to microstreaming and mixing, which accelerates cleaning, extraction, polymer degradation, and other processes (Yasuda and Koda 2011). In sonochemical equipment modeling, efforts have been focused on facilitating the chemistry rather than obtaining better acoustic parameters. The manufacturing of ultrasonic systems seems to accommodate the expanding usage of ultrasound in different applications. However, the main goal, which is a sonoreactor with high energy efficiency, is not yet industrially materialized (Toma et al. 2011).

For scaling up, it is necessary to first identify and understand the mechanism of interaction from observed ultrasonic phenomena in order to re-create the desired cavitation field on a larger scale. Then, optimum conditions to achieve the desired transformation in terms of operating/design variables that influence cavitation must then be established (Gogate et al. 2011). Various methods to examine the cavitational activity in sonoreactors have been proposed. The characterization of the cavitational phenomena and its effects are usually done through mapping, a stepwise procedure where quantification of cavitational activity in sonoreactor is done by means of primary effect (temperature or pressure measurement at the time of bubble collapse) and/or secondary effect (quantification of chemical or physical effects in terms of measurable quantities after bubble collapse) to identify the active and passive zones (Sutkar and Gogate 2009). The techniques employed for understanding the cavitational activity distribution is shown in Fig. 5.1.

To optimize an ultrasonic process, Mason and Cordemans (1998) recommended the following steps:

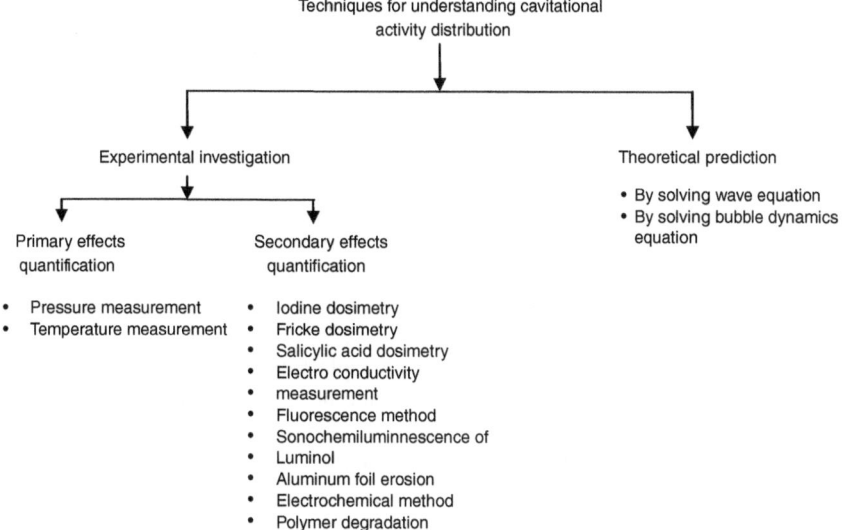

Fig. 5.1 Classification of different types of mapping techniques. Reprinted with permission from (Sutkar and Gogate 2009) Copyright (2009), Elsevier

1. Adding solids or gas bubbles to act as nuclei in order to ease cavitation process.
2. Try entraining different gases or mixture of gases.
3. Try using different kinds of solvents for different temperature ranges and cavitation energies.
4. Optimizing the power required for desired reaction.
5. Do not charge all components in the reactor at once when using a solid–liquid system.
6. If possible, try to homogenize the two-phase system as much as possible.
7. Try different shapes (diameter and volumes) for the reactor.
8. It would be preferred (but not always) to avoid standing wave conditions by performing sonochemical reactions under high power conditions with mechanical stirring.
9. If possible, try transforming a batch system into a continuous one.
10. Choose conditions which enable comparisons between different sonochemical reactions.

Bhirud et al. (2004) investigated a novel configuration of ultrasonic bath (equipped with transducer) with longitudinal vibrations for formic acid degradation (Fig. 5.2). A single longitudinally vibrating transducer was kept at the bottom of an 8 L holding capacity reactor (15 × 33 × 20 cm). The operating frequency of irradiation was 36 kHz with a 150 W of maximum power dissipation into the system. In comparison with other sonochemical reactors, this reactor gave about 4–5 times more cavitational yield than using multiple transducer irradiation

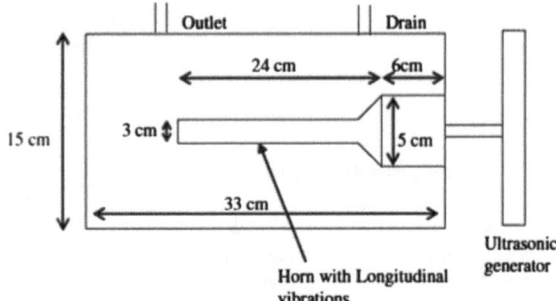

Fig. 5.2 Ultrasonic bath with longitudinal vibrating horn. Reprinted with permission from (Bhirud et al. 2004) Copyright (2009), Elsevier

(ultrasonic bath, dual frequency flow cell and triple frequency flow cell) and about twice the cavitational yield than using ultrasonic horn (single transducer localized irradiation). Further testing showed that this configuration could be potentially used for industrial-scale wastewater treatment applications (Bhirud et al. 2004). Cravotto et al. (2005) introduced several improvements on a conventional horn-type reactor operated at 18.2 kHz with a maximum power rating of 1,000 W (Fig. 5.3). To make sure that the reactor could be operated continuously at high intensities under stringent reaction conditions, a novel cooling system to control the temperature to below 40 °C and a more efficiently thermostatted reactor was introduced. A more uniform acoustic field and optimal acoustic streaming in every part of the reactor were achieved by rotating the reactor eccentrically around the horn axis with the probe moving alternatively up and down by a predetermined excursion at a chosen speed. The result was an increase in substitution index (% of alkylated glucosamine units) for formaldehyde from 60 to 86 % as compared to standard horn-type reactor in 3 h. This improvement was also observed for reductive amination of chitosan and a variety of reported aldehydes (Cravotto et al. 2005). Nikitenko et al. (2007) compared the performance of 20 kHz sonochemical reactors with different geometrical configurations using thermal probe method and two chemical dosimeters (using H_2O_2 and diphenylmethane). Results showed that the sonochemical reaction rates were driven by the total absorbed acoustic energy with little dependence on the geometry of the reactors. With an increase of horn surface, the sonochemical efficiency was enhanced at the same specific absorbed acoustic power due to the formation of larger bubble size (Nikitenko et al. 2007).

Although readily available, single transducer design does not work well in large-scale applications due to the fact that the cavitational activity mainly concentrates near the transducer (Kumar et al. 2007). Gogate and Pandit (2004) compared different cavitational equipments in terms of cavitational yields and energy efficiency and their results are summarized in Fig. 5.4. Among all the equipments, triple frequency flow cell was the most energy efficient due to its uniform energy dissipation over a wide area. It is not possible to conclude the applicability of multiple frequency irradiations for process intensification and synergistic effects because results are usually found to be dependent on the type of reaction (Gogate and Pandit 2004). Work done on multiple transducers was

Fig. 5.3 Modified horn-type reactor. (1) transducer and booster, (2) horn, (3) reaction tube, (4) reactor eccentric rotation, (5) vertical probe excursion, (6) cooling oil (chiller), (7) thermostatting fluid (peltier cells). Reprinted with permission from (Cravotto et al. 2005) Copyright (2005), Elsevier

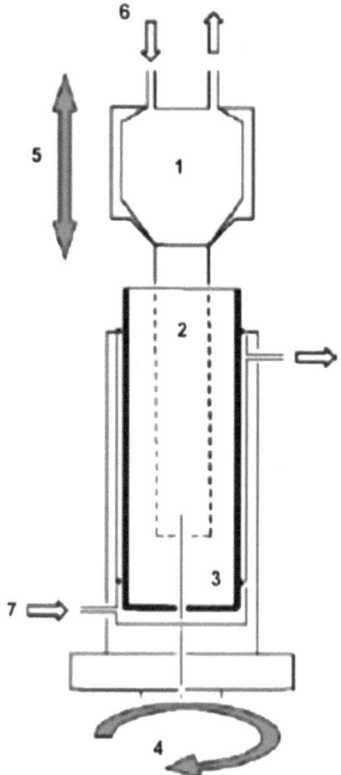

conducted by Kumar et al. (2007), who introduced two new sonochemical designs with a capacity of 7 L liquid. One design was based on a single large transducer located at the bottom of the reactor emitting radiation longitudinally away from the bottom perpendicular to the axis of the horn. Operating conditions and reactor geometry were similar to the one reported by Bhirud et al. (2004). Another reactor was a hexagonal flow cell, equipped with three transducers (20, 30 and 50 kHz) in a circular shape with diameter of 0.06 m and 0.03 spacing between each transducers as well as the bottom and top of the reactor (Fig. 5.5). The total maximum power that could be dissipated in this system was 900 W. Mapping of cavitational activity of the reactors by measuring the local pressure amplitude (using hydrophone) and cavitational activity (using cavitational activity indicator) revealed the near uniform distribution of cavitational activity. For both reactor designs, percentage variation in cavitational activity was found in the range of 10–30 % at different radial and axial locations, which was significantly lower than 80–400 % range using conventional immersion horn-type design over a small distance of radial and axial directions. The use of multiple frequencies also gave higher mean pressure amplitude and cavitational activity as compared to single frequency operation due to the greater dissipated power and higher intensity of cavity collapse (Kumar et al. 2007). Hodnett et al. (2007) presented a preliminary study on

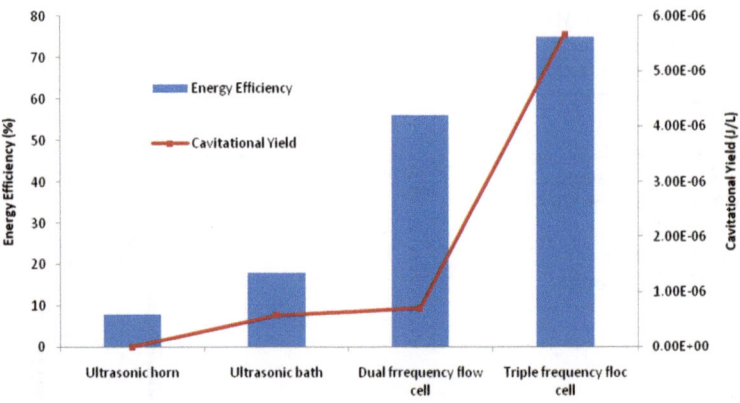

Fig. 5.4 Comparison between cavitational equipments in terms of energy efficiency and cavitational yield. Adapted with permission from Gogate and Pandit (2004)

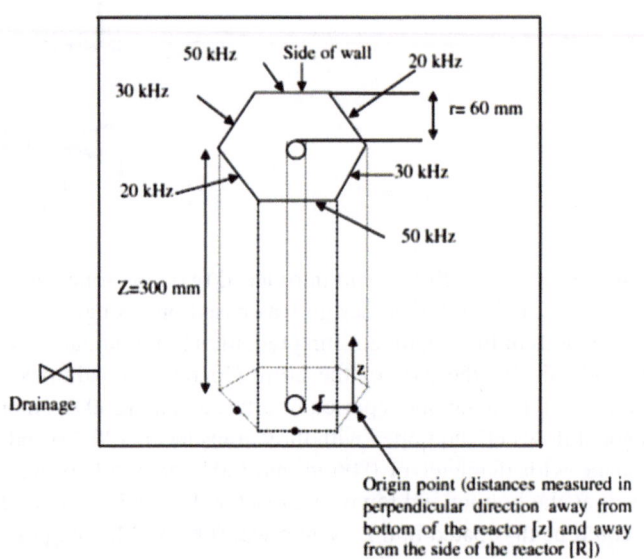

Fig. 5.5 Hexagonal flow cell. Reprinted with permission from (Kumar et al. 2007) Copyright (2007), Elsevier

the acoustic field distribution in a 25 kHz, 1.8 kW, 25 L (330 mm height, 312 mm internal diameter) cylindrical vessel. Thirty units of 25 kHz piezoelectric transducers were placed evenly in three horizontal rows of 10 devices around the vessel wall radiating into the central volume (Fig. 5.6). They found good evidence of reproducible acoustic performance, especially at power up to 100 W, where the standard deviation in the mean was 12 % (Hodnett et al. 2007).

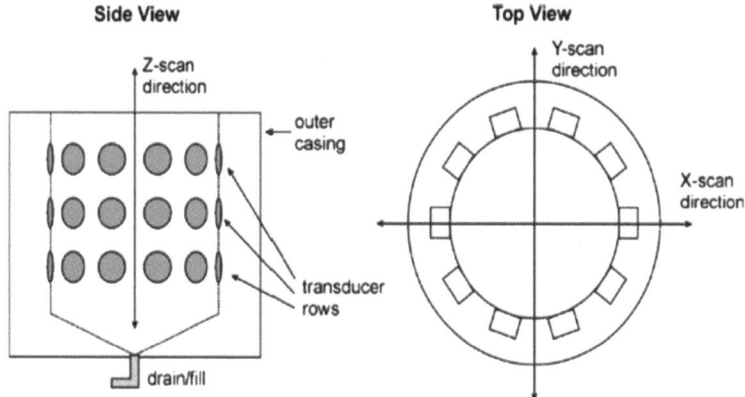

Fig. 5.6 Cylindrical sonoreactor with multiple transducers. Reprinted with permission from (Hodnett et al. 2007) Copyright (2007), Elsevier

Liu et al. (2008) used ultrasonic airlift loop reactor (ALR) which combined ultrasonic technique with ALR in the degradation of dimethoate (Fig. 5.7). The ALR consisted of coaxial tubes with working volume of 100 mL. Ultrasonic irradiation source (40 kHz, 0–240 W) was placed at the bottom of the reactor into the central column. Ozone was bubbled into the riser to improve circulation of the liquid in the reactor. Significant synergistic effect was observed by using this reactor because 90.8 % of the dimethoate was degraded after 4 h irradiation as compared to 20.1 and 14.5 % degradation by using ozone oxidation and ultrasound, respectively (Liu et al. 2008). Son et al. (2009) investigated the acoustic energy distribution for various frequencies (35, 72, 110 and 170 kHz) in a large-scale sonoreactor. In their study, an acrylic bath (L 1.2 m, W 0.6 m, H 0.4 m) was used with an ultrasonic transducer module, containing nine PZT transducers placed at the center of the side of the bath. At power input of 240 W, they showed that high frequency over 100 kHz could not propagate well in the large-scale sonoreactor with long-irradiation distance resulting in poor energy distribution. It was suggested that to optimize the use of high frequency ultrasound in sonoreactor with long-irradiation distance, larger energy input was required. The highest cavitation energy distribution was recorded in the case of ultrasonic irradiation at 72 kHz, while irradiation at 35 kHz showed larger half-cavitation-energy distance than in the case of 72 kHz (Son et al. 2009). de La Rochebrochard et al. (2012) presented an interesting study on individual and coupled effects of liquid height, frequency, and reactor configuration on sonochemical efficiency of a cup-horn sonoreactor. The effect of liquid height was found to be more dependent on reactor configuration (change in emitting-to-sonified surface ratio) as compared to frequency (20–500 kHz). The limits of acoustic zone in the reactor, which depended on both transducer diameter and frequency, significantly affected the production of radical species quantified by I_3^- formation rate (de La Rochebrochard et al. 2012).

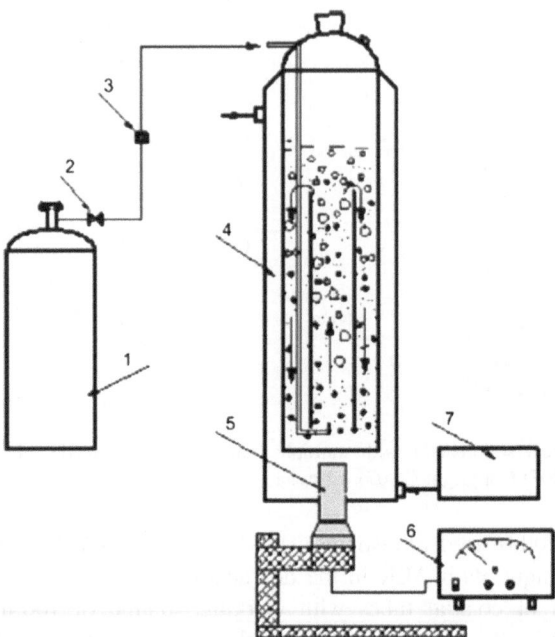

Fig. 5.7 Ultrasonic airlift loop reactor. (1) Gas cylinder; (2) stopvalve; (3) flowmeter; (4) airlift loop reactor; (5) ultrasonic probe; (6) ultrasonic generator; (7) thermostatic bath. Reprinted with permission from (Liu et al. 2008) Copyright (2008), Elsevier

One of the problems for sonoreactor design is that ultrasonic cavitation produces severe corrosion of the metallic vibrating surface and even vibrators of sonotrodes which are usually made of titanium alloy. As a result, contamination by the produced erosion products may occur (Dion 2009). Dion (2009) presented a new continuous sonoreactor based on cylindrical converging ultrasonic waves, which produced a powerful concentric and confined, chemically active cavitation zone in a tube, away from the wall in such a way that there was no erosion or contaminative products. These new sonoreactors exhibited industrial processing capabilities and measured in tonnes per hour for a 50 kHz model, depending on the cavitation energy per unit volume required to produce the desired effect (Dion 2009). Loranger et al. (2011) compared the sonochemical effects of a batchwise and a continuous ultrasonic system with the same transducer technology for the potential scale-up applications. Sonochemical effects were assessed using KI oxidation, while aluminium erosion was used to study the mechanical effects for both of the reactors. The usage of glass reactor was found to exhibit stronger effect on the attenuation of sonochemical effect as it acted as an ultrasonic resistance which reduced the sonochemical formation of I_3^- in this study. At 170 kHz and 1,000 W, sonochemical effect of a full-scale flow-through sonoreactor was 33 % more superior to ultrasonic bath for an irradiation time of 11.2 s per min of operation and 683 % better in terms of time corrected flow-through for reactor. Interestingly, full scale setup was more efficient as compared to laboratory unit under the same ultrasonic conditions obtained in this study (Loranger et al. 2011).

5.3 Conclusion

Although with rapid development of sonochemistry and considerable high potential for sonochemical reactors, there are still technical limitations and difficulties which prevent its wider use in industrial scale. Large-scale studies should be focused in maintaining uniform distribution of cavitational activity in sonochemical reactors by optimizing operation parameters (frequency of irradiation, intensity of irradiation and operating power dissipation per unit volume) and by proper design of the reactor itself (reactor configuration, location and number of transducers). Industrial sonoreactors should possess flexibility in terms of operating at different loadings of pollutants and should lead to economic savings. Future work must be focused on the development of large-scale multiple frequencies or multiple transducer reactors, which are able to operate in continuous mode. Without doubt, combined efforts from expertise in different fields such as chemists, physicists, chemical engineers, and equipment manufacturers are needed in order to truly exploit the potential of using ultrasonic irradiation as a "green" technology for environmental remediation.

References

Adewuyi YG (2005) Sonochemistry in environmental remediation. 1. Combinative and hybrid sonophotochemical oxidation processes for the treatment of pollutants in water. Environ Sci Technol 39:3409–3420

Bhirud US, Gogate PR, Wilhelm AM, Pandit AB (2004) Ultrasonic bath with longitudinal vibrations: a novel configuration for efficient wastewater treatment. Ultrason Sonochem 11:143–147

Bizzi CA, Müller EI, de Moraes Flores ÉM, Duarte FA, Korn M, Nunes MAG, Mello PA, Dressker VL (2011) Ultrasound-assisted industrial synthesis and processes. In: Chen D, Sharma SK, Mudhoo SK (eds) Handbook on application of ultrasound: sonochemistry for sustainability. CRC Press, Taylor & Francis Group, Boca Raton

Chen WS, Huang YL (2011) Removal of dinitrotoluenes and trinitrotoluene from industrial wastewater by ultrasound enhanced with titanium dioxide. Ultrason Sonochem 18:1232–1240

Cravotto G, Omiccioli G, Stevanato L (2005) An improved sonochemical reactor. Ultrason Sonochem 12:213–217

de La Rochebrochard S, Suptil J, Blais JF, Naffrechoux E (2012) Sonochemical efficiency dependence on liquid height and frequency in an improved sonochemical reactor. Ultrason Sonochem 19:280–285

Dion JL (2009) Contamination-free high capacity converging waves sonoreactors for the chemical industry. Ultrason Sonochem 16:212–220

Gallego-Juárez JA, Rodriguez G, Acosta V, Riera E (2010) Power ultrasonic transducers with extensive radiators for industrial processing. Ultrason Sonochem 17:953–964

Gogate PR (2007) Application of cavitational reactors for water disinfection: current status and path forward. J Environ Manage 85:801–815

Gogate PR, Pandit AB (2004) Sonochemical reactors: scale up aspects. Ultrason Sonochem 11:105–117

Gogate PR, Sutkar VS, Pandit AB (2011) Sonochemical reactors: important design and scale up considerations with a special emphasis on heterogeneous systems. Chem Eng J 166:1066–1082

Grčić I, Obradović M, Vujević D, Koprivanac N (2010) Sono-Fenton oxidation of formic acid/formate ions in an aqueous solution: from an experimental design to the mechanistic modeling. Chem Eng J 164:196–207

Hodnett M, Choi MJ, Zeqiri B (2007) Towards a reference ultrasonic cavitation vessel. Part 1: preliminary investigation of the acoustic field distribution in a 25 kHz cylindrical cell. Ultrason Sonochem 14:29–40

Katekhaya S, Gogate PR (2011) Intensification of cavitational activity in sonochemical reactors using different additives: efficacy assessment using a model reaction. Chem Eng Process 50:95–103

Kumar A, Gogate PR, Pandit AB (2007) Mapping the efficacy of new designs for large scale sonochemical reactors. Ultrason Sonochem 14:538–544

Liu Y-N, Jin D, Lu X-P, Han P-F (2008) Study on degradation of dimethoate solution in ultrasonic airlift loop reactor. Ultrason Sonochem 15:755–760

Loranger E, Paquin M, Daneault C, Chabot B (2011) Comparative study of sonochemical effects in an ultrasonic bath and in a large-scale flow-through sonoreactor. Chem Eng J 178:359–365

Madhavan J, Grieser F, Ashokkumar M (2010) Combined advanced oxidation processes for the synergistic degradation of ibuprofen in aqueous environments. J Hazard Mater 178:202–208

Mahamuni NN, Adewuyi YG (2010) Advanced oxidation processes (AOPs) involving ultrasound for waste water treatment: a review with emphasis on cost estimation. Ultrason Sonochem 17:990–1003

Mason TJ (1992) Industrial sonochemistry: potential and practicality. Ultrasonics 30:192–196

Mason TJ, Cordemans ED (1998) Practical considerations for process optimization. In: Luche JL, Bianchi C (eds) Synthetic organic sonochemistry. Plenum Publishers, New York

Naddeo V, Belgiorno V, Kassinos D, Mentzavinos D, Meric S (2010) Ultrasonic degradation, mineralization and detoxification of diclofenac in water: optimization of operating parameters. Ultrason Sonochem 17:179–185

Nikitenko SI, Le Naour C, Moisy P (2007) Comparative study of sonochemical reactors with different geometry using thermal and chemical probes. Ultrason Sonochem 14:330–336

Son Y, Lim M, Khim J (2009) Investigation of acoustic cavitation energy in a large-scale sonoreactor. Ultrason Sonochem 16:552–556

Sutkar VS, Gogate PR (2009) Design aspects of sonochemical reactors: techniques for understanding cavitational activity distribution and effect of operating parameters. Chem Eng J 155:26–36

Thangavadivel K, Megharaj M, Mudhoo A, Naidu R (2011) Degradation of organic pollutants using ultrasound. In: Chen D, Sharma SK, Mudhoo A (eds) Handbook on applications of ultrasound: sonochemistry for sustainability. CRC Press, Taylor & Francis Group, Boca Raton

Toma M, Fukutomi S, Asakura Y, Koda S (2011) A calorimetric study of energy conversion efficiency of a sonochemical reactor at 500 kHz for organic solvents. Ultrason Sonochem 18:197–208

Wang XK, Wei YC, Wang C, Guo WL, Wang JG, Jiang JX (2011) Ultrasonic degradation of reactive brilliant red K-2BP in water with CCl_4 enhancement: performance optimization and degradation mechanism. Sep Purif Technol 81:69–76

Yasuda K, Koda S (2011) Development of sonochemical reactor. In: Chen D, Sharma SK, Mudhoo A (eds) Handbook on application of ultrasound: sonochemistry for sustainability. CRC Press, Taylor & Francis Group, Boca Raton